Robert A. Beaudet
Seaver Science Center 622
University of Southern California
Los Angeles, California 90007

Basic Principles
of Spectroscopy

Basic Principles of Spectroscopy

Raymond Chang
Williams College

McGraw-Hill Book Company

New York   St. Louis   San Francisco   Düsseldorf
Johannesburg   Kuala Lumpur   London   Mexico   Montreal
New Delhi   Panama   Rio de Janeiro   Singapore   Sydney   Toronto

Basic Principles of Spectroscopy

Library of Congress Catalog Card Number 74-132340

07-010517-0

2 3 4 5 6 7 8 9 0   M A M M   7 9 8 7 6 5 4 3 2

This book was set in Modern by The Maple Press Company,
and printed on permanent paper and bound by The Maple
Press Company. The drawings were done by John Cordes,
J. & R. Technical Services, Inc. The editors were James L.
Smith and Madelaine Eichberg. Matt Martino supervised
production.

Preface

Spectroscopy is now taught at the undergraduate level in most schools; however, the introduction of the subject is varied and sometimes unsatisfactory. For example, an undergraduate's usual encounter with the subject is either through a quantum mechanics course (atomic, infrared, Raman, and ultraviolet spectroscopy) or an instrumental analysis course (atomic, infrared, and ultraviolet spectroscopy). Topics such as infrared, ultraviolet, and nuclear magnetic resonance are also covered in biochemistry and organic chemistry. As a result the student may be overexposed in some areas and know very little about others. Also, there is always the added confusion when these topics are presented with different approaches and emphases. Since the basic principles of all branches of spectroscopy are essentially the same, there is a distinct advantage in teaching the subject as a whole. The student should better understand the fundamental concepts when he acquires an overall perspective.

The topics chosen in this text include the most common branches. Chapters 1 and 2, which discuss elementary quantum mechanics and group theory, form the vocabulary for spectroscopy. The emphasis in each subsequent chapter is to present the most basic ideas and to illustrate, with simple examples, the applications and limitations of each technique. Description of the experimental set up is kept at a minimum. In attempting to cover such a wide range of materials and to keep the text within reasonable length, it has not been possible to include more discussions for in-depth understanding of each branch. Thus this text should be more suitable for those who wish to be acquainted with the general aspects of spectroscopy rather than specialize in two or three areas. References to standard texts are listed in the back of each chapter. Particular attention should be given to the Reading Assignments, which contain references to articles largely drawn from the *Journal of Chemical Education*. These articles are written at the introductory level and therefore should supplement the text in both theory and application. The student is strongly urged to read them carefully.

This book is written as a one-semester course for advanced undergraduates and beginning graduate students in chemistry and biochemistry who have previously taken a standard course in physical chemistry.

I wish to thank my former teacher Dr. Charles Johnson, Jr. for introducing me to spectroscopy and Dr. Charles Compton for his interest and encouragement. I am also grateful to Dr. William Moomaw for his very helpful comments on the first six chapters and for bringing to my attention the recently developed photoelectron spectroscopy technique. Finally, I wish to express my appreciation to Dr. Eugene Olsen for his careful and critical reading of the entire manuscript.

Raymond Chang

Contents

1
Basic Principles

1-1 INTRODUCTION

Spectroscopy is the study of the interaction of electromagnetic radiation with matter. Detailed information regarding molecular structure (molecular symmetry, bond distances, and bond angles) and chemical properties (electronic distribution, bond strength, and intra- and inter-molecular processes) can be obtained from the atomic and molecular spectra.

Maxwell (1855) first pointed out that visible light is just one form of electromagnetic waves, the complete spectrum of which includes microwaves, infrared, x-rays, and γ-rays. An electromagnetic wave of any kind consists of an electric field component ϵ and a magnetic field component H as shown in Fig. 1-1. For electromagnetic waves traveling in the x direction, we write

$$\epsilon = \epsilon_0 \sin 2\pi \left(\frac{x}{\lambda} - \nu t \right) \tag{1-1}$$

$$H = H_0 \sin 2\pi \left(\frac{x}{\lambda} - \nu t \right) \tag{1-2}$$

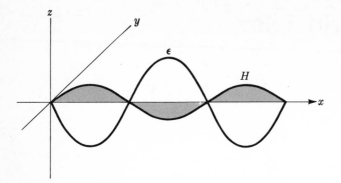

Fig. 1-1 The electric and magnetic field components of a plane-polarized wave.

where λ and ν are the wavelength and frequency of the radiation. The velocity c of the wave is given by

$$c = \lambda \nu \qquad (1\text{-}3)$$

Thus we can think of an electromagnetic wave as consisting of an electric and a magnetic field oscillating in space, the direction of oscillation being perpendicular to the direction of wave propagation. The interaction of electromagnetic radiation with matter is either between the electric field component of the wave and the electric properties or between the magnetic field component and the magnetic properties of the atoms and molecules.

Although spectroscopic investigations were first carried out by Newton in the seventeenth century, quantitative treatment of the subject was possible only after the introduction of quantum mechanics. Indeed, it was atomic spectra that provided the first direct experimental evidence for the various quantum mechanical postulates. Therefore, at least a qualitative understanding of quantum mechanics is required to explain the spectroscopic observations, and we shall review some of the basic principles in the next section.

1-2 REVIEW OF QUANTUM MECHANICS

Toward the end of the nineteenth century it was becoming increasingly evident that many of the physical laws which had been applied in classical mechanics could not find similar success when applied to systems on the atomic and molecular scale. The first breakthrough came in 1900 when Planck, in his study of blackbody radiation, found that the laws of radiation of energy could not be explained on the basis of thermodynamics, and was consequently led to the hypothesis that absorption and emission of radiation energy by matter do not take place continuously but in finite *quanta* of energy. The energy E is proportional to the frequency ν of the harmonic oscillator responsible for the absorption and emission of radiation, and we have

$$E = h\nu \tag{1-4}$$

where h is the Planck constant.

The idea of *quantization of energy* was soon adopted by Einstein (1905) to explain the photoelectric effect. It was noticed that when light of a certain wavelength fell on a metal surface in a vacuum, the surface became positively charged and electrons were ejected. For light of wavelength greater than a certain value, no such effect was observed; below this value the number of electrons ejected was proportional to the intensity of the light. However, the velocity of the ejected electrons depended only on the frequency of the light and not on its intensity. Einstein suggested that light consists of quanta of energy $h\nu$, called *photons*, which travel with the velocity of light. Thus when a photon of energy $h\nu$ strikes the surface, an electron is ejected with velocity v, and the kinetic energy of the electron is given by

$$\tfrac{1}{2}m_e v^2 = h\nu - W \tag{1-5}$$

where m_e is the electronic mass and W is a constant characteristic of the metal.

In 1913 Bohr made the great advance by applying the quantum theory to the study of atomic spectra. The emission spectra of hydrogen atoms were found to consist of a number of sharp lines, the separation between which could be fitted into a mathematical series. In order to explain the fact that atomic hydrogen emits only certain characteristic frequencies, Bohr postulated that in atoms electrons occupy states or levels of certain definite energy values. The emission of light is caused by the electron falling from a higher (E_h) to a lower (E_l) energy state. We write

$$E_h - E_l = \Delta E = h\nu \tag{1-6}$$

One photon of energy $h\nu$ is emitted when such an electronic transition occurs. Conversely, the atom can absorb a photon upon going from the lower to the higher electronic state. Equation (1-6) forms the basis of all quantitative studies of spectroscopy (Fig. 1-2).

Applying Planck's quantum theory and using Newton's laws of motion, Bohr was able to obtain the correct formula for the hydrogen electronic energy levels and thus explain the appearance of the spectral lines. However, Bohr's theory had to be limited by certain arbitrary restrictions, and more refined experiments showed that his postulates

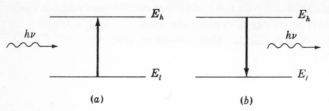

Fig. 1-2 Interaction of electromagnetic radiation with matter.
(*a*) Absorption. (*b*) Emission.

were not adequate to account for all the observed phenomena.   Up to this point (1916), the quantum theory had enjoyed a fruitful but limited success in spectroscopy.

The inadequacy of the old quantum theory, as it is now called, lies in the fact that when we are dealing with systems on the atomic scale we can no longer apply the laws of classical mechanics.   A new concept and new mathematical equations are required to correlate the behavior of these minute objects, and this then leads us to the advent of quantum mechanics.

The work of Planck, Einstein, and Bohr presented a dilemma with respect to the nature of light.   The postulate that light consists of photons means a *particle* concept of nature.   On the other hand, the phenomena of diffraction and interference can be explained only in terms of a *wave* theory.   Unfortunately, there is no simple resolution to this duality. We have been accustomed to think in terms of macroscopic systems, and therefore we expect every system to have a set of well-defined properties. But there is no reason that this assumption must also apply to microscopic systems.   Thus, it is not unreasonable to think of light as possessing both wave and particle properties—with Maxwell's theory describing the propagation of light and the quantum theory describing the interaction of light with matter.   In 1924 de Broglie suggested that the particle property of matter, that is, the momentum p, is related to its wave property λ by the Planck constant.   Hence we have

$$p = mv = \frac{h}{\lambda} \tag{1-7}$$

A number of experiments have indeed confirmed the dual nature of electrons and photons.

Another very important fundamental quantum mechanical postulate was given by Heisenberg (1927), who stated that it is impossible to determine precisely both conjugate variables of a physical system at the same time.   Examples of conjugate variables in classical mechanics are energy and time, position and momentum, angle and angular momentum, etc.

According to the Heisenberg uncertainty principle, the uncertainty in the simultaneous determination of any two conjugates X and Y is

$$\Delta X \, \Delta Y \sim \hbar \qquad (1\text{-}8)$$

where $\hbar$, or "h bar," is h divided by 2π. We therefore write

$$\Delta E \, \Delta t \sim \hbar \qquad (1\text{-}9)$$

and

$$\Delta p \, \Delta q \sim \hbar \qquad (1\text{-}10)$$

where p and q denote momentum and position.

Since newtonian laws could not explain any of the experiments cited earlier, it became apparent that a new form of mechanics would have to be introduced to describe the behavior of the microscopic system. An equation of motion for these systems must incorporate both particle and wave properties, and such an equation was given by Schrödinger in 1926.

In classical mechanics the total energy E of a conservative system is equal to its kinetic energy T and potential energy V. Hence

$$E = T + V \qquad (1\text{-}11)$$

For a particle of mass m moving with velocity v along the x axis in a potential field $V(x)$, we write

$$T = \tfrac{1}{2}m\left(\frac{dx}{dt}\right)^2 \qquad (1\text{-}12)$$

$$E = \tfrac{1}{2}m\left(\frac{dx}{dt}\right)^2 + V(x) = \frac{p^2}{2m} + V(x) \qquad (1\text{-}13)$$

where

$$v = \frac{dx}{dt} \qquad (1\text{-}14)$$

The conversion of Eq. (1-13) into Schrödinger's wave equation is carried out by replacing momentum p with an operator $(\hbar/i)(d/dx)$, where $i = \sqrt{-1}$. Thus

$$p \to \frac{\hbar}{i}\frac{d}{dx}$$

$$p^2 \to \left(\frac{\hbar}{i}\frac{d}{dx}\right)^2 = -\hbar^2\frac{d^2}{dx^2}$$

and Eq. (1-13) becomes

$$-\frac{\hbar^2}{2m}\frac{d^2}{dx^2} + V(x) = E \qquad (1\text{-}15)$$

We note that $(-\hbar^2/2m)(d^2/dx^2)$ is only an operator, that is, it is a mathematical symbol that tells us specifically what to do to a function ψ. Equation (1-15) is an operator equation, the complete form of which must be written as

$$-\frac{\hbar^2}{2m}\frac{d^2\psi}{dx^2} + V\psi = E\psi \tag{1-16}$$

Equation (1-16) is the time-independent Schrödinger equation in one dimension if ψ is the wave function which describes the wave property of the particle with mass m. In three dimensions we write

$$\nabla^2\psi + \frac{2m}{\hbar^2}(E - V)\psi = 0 \tag{1-17}$$

where ∇^2, the laplacian operator, is given by

$$\nabla^2 \equiv \frac{\partial^2}{\partial x^2} + \frac{\partial^2}{\partial y^2} + \frac{\partial^2}{\partial z^2}$$

Rearrangement of Eq. (1-17) gives

$$\left(-\frac{\hbar^2}{2m}\nabla^2 + V\right)\psi = E\psi$$

or

$$\mathcal{H}\psi = E\psi \tag{1-18}$$

where $\mathcal{H}$, the hamiltonian operator, is given by

$$\mathcal{H} = -\frac{\hbar^2}{2m}\nabla^2 + V$$

A more fundamental wave function must also include the time-dependent part, and when this is done our wave equation becomes

$$\mathcal{H}\psi(q,t) = i\hbar\frac{\partial\psi(q,t)}{\partial t} \tag{1-19}$$

This is the time-dependent Schrödinger equation. The classical dynamic variable, that is, the observable quantity E, is converted into the quantum mechanical operator $i\hbar(\partial/\partial t)$ and ψ is replaced by $\psi(q,t)$, where q denotes the coordinates. To be physically meaningful, $\psi(q,t)$ must have the following properties:

1. $\psi(q,t)$ is single-valued and finite.
2. $\psi(q,t)$ and $\partial\psi(q,t)/\partial q$ are continuous functions.
3. $\psi(q,t)$ is normalized, that is,

$$\int_{-\infty}^{+\infty}\psi^*(q,t)\psi(q,t)\,dq = 1$$

where $\psi^*(q,t)$ is the complex conjugate of $\psi(q,t)$.

4. In general, an observable quantity α has an associated wave mechanical operator α_{op} such that

$$\alpha_{op}\psi(q,t) = \alpha\psi(q,t)$$

and the average value of α, $\bar{\alpha}$, is given by

$$\bar{\alpha} = \langle\alpha\rangle = \int_{-\infty}^{+\infty} \psi^*(q,t)\alpha_{op}\psi(q,t)\,dq = \lim_{N\to\infty}\frac{1}{N}\sum_i^N \alpha_i$$

where N is the number of observations made. $\psi(q,t)$ can be written as a product of two functions f and ϕ:

$$\psi(q,t) = f(q)\phi(t) \tag{1-20}$$

Substituting Eq. (1-20) into (1-19) we get

$$\frac{1}{f(q)}\,\mathcal{K}f(q) = i\hbar\,\frac{1}{\phi(t)}\,\frac{\partial\phi(t)}{\partial t} \tag{1-21}$$

Since q and t are independent variables, each side of Eq. (1-21) can be set equal to a constant C. Hence

$$\frac{1}{f(q)}\,\mathcal{K}f(q) = C \tag{1-22}$$

and

$$\frac{i\hbar}{\phi(t)}\,\frac{\partial\phi(t)}{\partial t} = C \tag{1-23}$$

Obviously the constant C is just E [see Eq. (1-18)]. Thus

$$\frac{i\hbar}{\phi(t)}\,\frac{\partial\phi(t)}{\partial t} = E \tag{1-24}$$

The solution of Eq. (1-24) gives

$$\phi(t) = e^{-iEt/\hbar} \tag{1-25}$$

The total wave function $\psi(q,t)$ is given by

$$\psi(q,t) = f(q)e^{-iEt/\hbar} \tag{1-26}$$

Depending on the particular situation, we shall use either the time-dependent or time-independent Schrödinger equation.

We are now in the position to apply the Schrödinger equation to a number of systems of interest in spectroscopic studies. In particular, we shall discuss the following four situations: The particle in a box, the hydrogen atom, the rigid rotator, and the linear harmonic oscillator.

THE PARTICLE IN A BOX

This is only an imaginary problem but the results are useful in illustrating the basic concepts of quantum theory. Consider the situation in which a particle of mass m is bound by two infinite potential walls (Fig. 1-3). Inside the box the potential energy V of the particle is zero. The time-independent Schrödinger equation for the particle moving along the x axis inside the box is

$$-\frac{\hbar^2}{2m}\frac{d^2\psi}{dx^2} = E\psi \tag{1-27}$$

Outside the box the particle would acquire an infinite amount of energy, which is clearly impossible. To obtain an explicit expression for E we must choose a certain function for ψ in Eq. (1-27). Equation (1-27) shows that the function ψ, when differentiated twice, gives the same function multiplied by some number E. Examples of these are the trigonometric functions

$$\frac{d^2(\sin ax)}{dx^2} = -a^2 \sin ax \qquad \frac{d^2(\cos ax)}{dx^2} = -a^2 \cos ax$$

Thus as a trial solution to Eq. (1-27) we choose the function

$$\psi = A \sin ax + B \cos ax \tag{1-28}$$

Substituting Eq. (1-28) into (1-27) we obtain

$$\frac{\hbar^2 a^2}{2m}\psi = E\psi$$

or

$$E = \frac{a^2 \hbar^2}{2m} \tag{1-29}$$

What is a? This can be obtained by applying the *boundary conditions* as follows: At the wall of the box the potential is infinite so there is no probability of finding the particle; therefore, at $x = 0$, $\psi = 0$. When

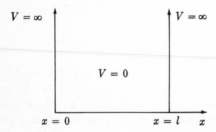

Fig. 1-3 A one-dimensional box with infinite potential walls.

this condition is applied to Eq. (1-28), we obtain

$$0 = B \qquad (1\text{-}30)$$

Similarly, at $x = l$, $\psi = 0$, and Eq. (1-28) becomes

$$\psi = A \sin al = 0$$

This means that

$$al = n\pi$$

where $n = 0, 1, 2, \ldots$, or

$$a = \frac{n\pi}{l}$$

Thus, in general, Eq. (1-28) can be written as

$$\psi_n = A \sin \frac{n\pi}{l} x \qquad (1\text{-}31)$$

where $n = 1, 2, 3, \ldots$, and the corresponding energies are

$$E_n = \frac{n^2 h^2}{8ml^2} \qquad (1\text{-}32)$$

We note first that according to Eq. (1-32) the energy of the particle does not vary continuously; rather, it can have only values of $h^2/8ml^2$, $h^2/2ml^2$, $9h^2/8ml^2$, $\ldots$.

Also, the system has a nonzero lowest-energy level. This can be explained by the Heisenberg uncertainty principle as follows: Since the particle is somewhere between $x = 0$ and $x = l$, there is a finite, nonzero uncertainty in determining its position. According to Eq. (1-10), therefore, there must also be a finite, nonzero uncertainty in determining its momentum and hence energy. Therefore, the particle has a *zero-point* energy. According to classical mechanics the probability of finding the particle is the same anywhere along the box, but this is not so for a quantum mechanical system. The wave function ψ has no physical meaning by itself and in fact may even be imaginary. On the other hand, the product $\psi^*\psi$ is necessarily real and may be interpreted as

$$\psi^*\psi \, dx = \text{probability that the system will be found between } x \text{ and}$$
$$x + dx$$

Since the particle must be in the box, we write

$$\int_0^l \psi^*\psi \, dx = 1$$

or

$$\int_0^l \left(A \sin \frac{n\pi}{l} x \right)^2 dx = 1$$

Hence

$$A^2 = \frac{2}{a}$$

or

$$A = \sqrt{\frac{2}{a}}$$

Thus

$$\psi_n = \sqrt{\frac{2}{a}} \sin \frac{n\pi}{l} x \qquad (1\text{-}33)$$

The wave function is said to be *normalized*. Figure 1-4 shows the plots of the energy levels and the corresponding ψ's and ψ^2's.

Finally we note that as m approaches the mass of a macroscopic system, the spacing between successive levels will approach the point where the energy will vary continuously.

THE HYDROGEN ATOM

The particle-in-a-box problem, although an imaginary one, is nevertheless useful since it serves to illustrate several basic quantum ideas and introduces the quantum number n. We shall now apply the Schrödinger equation to the hydrogen atom—the simplest atomic system.

This is a two-body problem since in the hydrogen atom we have an electron and a nucleus. The potential energy due to the electrostatic interaction is $-e^2/r$, where r is the distance between the electron and the nucleus. We write

$$\nabla^2\psi + \frac{2\mu}{\hbar^2}\left(E + \frac{e^2}{r}\right)\psi = 0 \qquad (1\text{-}34)$$

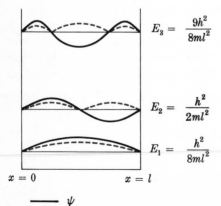

$$E_3 = \frac{9h^2}{8ml^2}$$

$$E_2 = \frac{h^2}{2ml^2}$$

$$E_1 = \frac{h^2}{8ml^2}$$

$x = 0$ $x = l$

——— ψ

– – – ψ^2

Fig. 1-4 Plots of ψ, ψ^2, and the energy levels for the particle in a one-dimensional box.

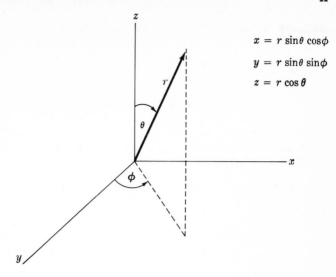

Fig. 1-5 Spherical polar coordinates.

where μ, the reduced mass, is defined in terms of the electronic mass m_e and the proton mass M_P as

$$\frac{1}{\mu} = \frac{1}{m_e} + \frac{1}{M_P} \tag{1-35}$$

Hence

$$\mu = \frac{m_e M_P}{m_e + M_P}$$

The advantage of using the reduced mass is that we can now treat the problem as if the system contained only a single particle of mass μ. Equation (1-34) is written in the cartesian coordinates. However, for central-force problems (that is, the force of attraction is spherically symmetric) it is often convenient to use the polar coordinates (Fig. 1-5). In polar coordinates, Eq. (1-34) becomes

$$\frac{\partial^2 \psi}{\partial r^2} + \frac{2}{r}\frac{\partial \psi}{\partial r} + \frac{1}{r^2 \sin\theta}\frac{\partial[\sin\theta\,(\partial\psi/\partial\theta)]}{\partial\theta} + \frac{1}{r^2 \sin\theta}\frac{\partial^2\psi}{\partial\phi^2}$$
$$+ \frac{2\mu}{\hbar^2}\left(E + \frac{e^2}{r}\right)\psi = 0 \tag{1-36}$$

The total wave function ψ can be written as a product of the radial part R and angular part Y as

$$\psi = R(r)Y_{l,m}(\theta,\phi) \tag{1-37}$$

Substituting Eq. (1-37) into (1-36), we obtain

$$\frac{1}{R}\frac{\partial[r^2(\partial R/\partial r)]}{\partial r} + \frac{2\mu}{\hbar^2}\left(E + \frac{e^2}{r}\right)$$
$$= -\frac{1}{Y_{l,m}\sin\theta}\frac{\partial[\sin\theta(\partial Y_{l,m}/\partial\theta)]}{\partial\theta} - \frac{1}{Y_{l,m}\sin^2\theta}\frac{\partial^2 Y_{l,m}}{\partial\phi^2} \quad (1\text{-}38)$$

Since the left-hand side of Eq. (1-38) is a function of r only, and the right-hand side a function of θ and ϕ, we can set each side equal to a constant $l(l + 1)$. Thus

$$\frac{d^2R}{dr^2} + \frac{2}{r}\frac{dR}{dr} + \left[\frac{2\mu}{\hbar^2}\left(E + \frac{e^2}{r}\right) - \frac{l(l+1)}{r^2}\right]R = 0 \quad (1\text{-}39)$$

and

$$\frac{1}{\sin\theta}\frac{\partial[\sin\theta(\partial Y_{l,m}/\partial\theta)]}{\partial\theta} + \frac{1}{\sin^2\theta}\frac{\partial^2 Y_{l,m}}{\partial\phi^2} + l(l+1)Y_{l,m} = 0 \quad (1\text{-}40)$$

The solutions of Eqs. (1-39) and (1-40) are given in the standard texts on quantum mechanics. Here we shall simply quote the results. The energy E, which appears only in Eq. (1-39), is given by

$$E_n = \frac{-\mu e^4}{2n^2\hbar^2} \quad (1\text{-}41)$$

where n, the principal quantum number, has the values of $n = 1, 2, 3, \ldots$. The radial function is given by

$$R_{n,l}(\rho) = \rho^l e^{-\rho/2}L_{n+1}^{2l+1}(\rho) \quad (1\text{-}42)$$

where ρ is given by

$$\rho = \frac{2\mu e^2}{n\hbar^2}r \quad (1\text{-}43)$$

and L_{n+1}^{2l+1} is the associated Laguerre polynomial.

The function $Y_{l,m}$, which appears in Eq. (1-40), is called the *spherical harmonic*. The azimuthal quantum number l has the values $l = 0, 1, 2, \ldots$. For a given value of l there are $2l + 1$ values of m, the magnetic quantum number. The relations between these three quantum numbers are

$$l \leq n - 1 \quad \text{and} \quad -l \leq m \leq l$$

The value of the principal quantum number determines the energy, and the value of l determines the shape of the atomic orbitals. Table 1-1 shows the values of the first few hydrogen atomic-orbital wave functions. Plots of some $R_{n,l}$ and $Y_{l,m}(\theta,\phi)$ are shown in Figs. 1-6 and 1-7.

Table 1-1   Normalized atomic-orbital wave functions for hydrogen atom†

n	l	m	Orbital	$R_{n,l}(r)$	$Y_{l,m}(\theta,\phi)$
1	0	0	$1s$	$2\left(\dfrac{z}{a_0}\right)^{\frac{3}{2}} e^{-\rho}$	$\dfrac{1}{\sqrt{4\pi}}$
2	0	0	$2s$	$\dfrac{1}{\sqrt{8}}\left(\dfrac{z}{a_0}\right)^{\frac{3}{2}}(2-\rho)e^{-\rho/2}$	$\dfrac{1}{\sqrt{4\pi}}$
2	1	0	$2p_z$	$\dfrac{1}{\sqrt{24}}\left(\dfrac{z}{a_0}\right)^{\frac{3}{2}}\rho e^{-\rho/2}$	$\dfrac{3}{4\pi}\cos\theta$
2	1	1	$2p_x$	$\dfrac{1}{\sqrt{24}}\left(\dfrac{z}{a_0}\right)^{\frac{3}{2}}\rho e^{-\rho/2}$	$\dfrac{3}{4\pi}\sin\theta\cos\phi$
2	1	−1	$2p_y$	$\dfrac{1}{\sqrt{24}}\left(\dfrac{z}{a_0}\right)^{\frac{3}{2}}\rho e^{-\rho/2}$	$\dfrac{3}{4\pi}\sin\theta\sin\phi$
3	0	0	$3s$	$\dfrac{2}{81\sqrt{3}}\left(\dfrac{z}{a_0}\right)^{\frac{3}{2}}(27-18\rho+2\rho^2)e^{-\rho/3}$	$\dfrac{1}{\sqrt{4\pi}}$
3	1	0	$3p_z$	$\dfrac{4}{81\sqrt{6}}\left(\dfrac{z}{a_0}\right)^{\frac{3}{2}}(6\rho-\rho^2)e^{-\rho/3}$	$\sqrt{\dfrac{3}{4\pi}}\cos\theta$
3	1	1	$3p_x$	$\dfrac{4}{81\sqrt{6}}\left(\dfrac{z}{a_0}\right)^{\frac{3}{2}}(6\rho-\rho^2)e^{-\rho/3}$	$\sqrt{\dfrac{3}{4\pi}}\sin\theta\cos\phi$
3	1	−1	$3p_y$	$\dfrac{4}{81\sqrt{6}}\left(\dfrac{z}{a_0}\right)^{\frac{3}{2}}(6\rho-\rho^2)e^{-\rho/3}$	$\sqrt{\dfrac{3}{4\pi}}\sin\theta\sin\phi$
3	2	0	$3d_{z^2}$	$\dfrac{4}{81\sqrt{30}}\left(\dfrac{z}{a_0}\right)^{\frac{3}{2}}\rho^2 e^{-\rho/3}$	$\sqrt{\dfrac{5}{16\pi}}(3\cos^2\theta-1)$
3	2	1	$3d_{xz}$	$\dfrac{4}{81\sqrt{30}}\left(\dfrac{z}{a_0}\right)^{\frac{3}{2}}\rho^2 e^{-\rho/3}$	$\sqrt{\dfrac{15}{8\pi}}\sin\theta\cos\theta\cos\phi$
3	2	−1	$3d_{yz}$	$\dfrac{4}{81\sqrt{30}}\left(\dfrac{z}{a_0}\right)^{\frac{3}{2}}\rho^2 e^{-\rho/3}$	$\sqrt{\dfrac{15}{8\pi}}\sin\theta\cos\theta\sin\phi$
3	2	2	$3d_{x^2-y^2}$	$\dfrac{4}{81\sqrt{30}}\left(\dfrac{z}{a_0}\right)^{\frac{3}{2}}\rho^2 e^{-\rho/3}$	$\sqrt{\dfrac{15}{16\pi}}\sin^2\theta\cos 2\phi$
3	2	−2	$3d_{xy}$	$\dfrac{4}{81\sqrt{30}}\left(\dfrac{z}{a_0}\right)^{\frac{3}{2}}\rho^2 e^{-\rho/3}$	$\sqrt{\dfrac{15}{16\pi}}\sin^2\theta\sin 2\phi$

† Z = nuclear charge; $a_0 = \hbar^2/\mu e^2$.

THE RIGID ROTATOR

The simplest rigid rotator consists of two point masses joined by a rigid bar (Fig. 1-8).   The center of mass is at o so that

$$m_1 r_1 = m_2 r_2 \tag{1-44}$$

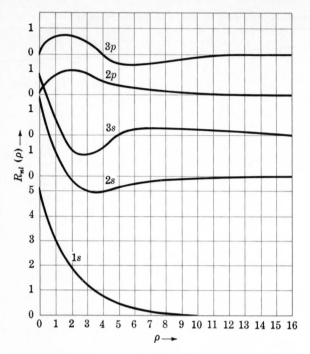

Fig. 1-6 Hydrogen atom radial wave functions for $1s$, $2s$, $2p$, $3s$, and $3p$ orbitals shown as functions of the variable $\rho = 2r/n$ (r in units of a_0). (*By permission of F. L. Pilar, "Elementary Quantum Chemistry," McGraw-Hill Book Company, New York, 1968.*)

According to classical mechanics, the treatment of circular motion requires the transformation of the linear system into the angular one. We write

$$
\begin{aligned}
\text{Coordinate } x &\rightarrow \text{angle } \phi \\
\text{Mass } m &\rightarrow \text{moment of inertia } I \\
\text{Velocity } v &\rightarrow \text{angular velocity } \omega \\
\text{Momentum } p &\rightarrow \text{angular momentum } p_\phi \ (= I\omega) \\
\text{Kinetic energy } \frac{p^2}{2m} &\rightarrow \frac{p_\phi{}^2}{2I} \\
\text{Potential energy } V(x) &\rightarrow V(\phi)
\end{aligned}
$$

The moment of inertia I is given by

$$
I = m_1 r_1{}^2 + m_2 r_2{}^2 \tag{1-45}
$$

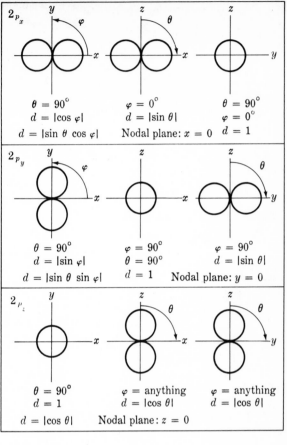

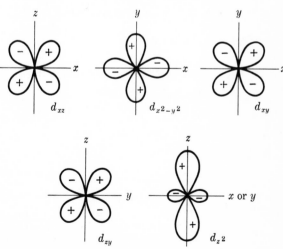

Fig. 1-7 Polar plots of $d = h(r)|Y_{l,m}(\theta,\phi)|$ for $h(r) =$ constant. (*By permission of F. L. Pilar, "Elementary Quantum Chemistry," McGraw-Hill Book Company, New York, 1968.*)

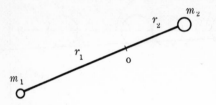

Fig. 1-8 A rigid rotator.

Since

$$r_1 = \frac{m_2}{m_1} r_2 = \frac{m_2}{m_1} (r - r_1) \qquad r_2 = \frac{m_1}{m_2} r_1 = \frac{m_1}{m_2} (r - r_2)$$

where $r = r_1 + r_2$, we obtain

$$r_1 = \frac{m_2}{m_1 + m_2} r \qquad (1\text{-}46)$$

and

$$r_2 = \frac{m_1}{m_1 + m_2} r \qquad (1\text{-}47)$$

Equation (1-45) can now be written as

$$I = \frac{m_1 m_2}{m_1 + m_2} r^2 = \mu r^2 \qquad (1\text{-}48)$$

where μ is the reduced mass.

If the rotation is unrestricted, the potential energy $V(\phi)$ becomes a constant and we can arbitrarily choose $V(\phi) = 0$. The time-independent Schrödinger equation for the rigid rotator is then given by

$$\nabla^2 \psi + \frac{2IE}{\hbar^2} \psi = 0 \qquad (1\text{-}49)$$

where the laplacian operator is given in terms of polar coordinates as

$$\nabla^2 = \frac{1}{r^2 \sin \theta} \left\{ \frac{\partial[\sin \theta(\partial/\partial\theta)]}{\partial\theta} + \frac{1}{\sin \theta} \frac{\partial^2}{\partial\phi^2} \right\}$$

The solution of Eq. (1-49) gives

$$E = \frac{\hbar^2}{2I} J(J + 1) \qquad (1\text{-}50)$$

where J, the rotational quantum number, is given by $J = 0, 1, 2, \ldots$. Classically, the energy for a rigid rotator is simply

$$E = \tfrac{1}{2} I \omega^2 \qquad (1\text{-}51)$$

The wave function ψ^{JM} is given by

$$\psi^{JM} = \Phi_M(\phi) \Theta_{MJ}(\theta)$$

where M is the component of J in a specially chosen direction and for a given value of J there are $2J + 1$ values of M. The degeneracy is thus $2J + 1$ since the energy of the rotator E does not depend on M in the absence of an electric or magnetic field. The first few wave functions are:

$$\psi^{00} = \frac{1}{\sqrt{2\pi}}$$

$$\psi^{10} = \sqrt{\frac{3}{4\pi}} \cos \theta$$

$$\psi^{1\pm1} = \sqrt{\frac{3}{8\pi}} \sin \theta (e^{\pm i\theta})$$

$$\psi^{20} = \sqrt{\frac{5}{16\pi}} (3 \cos^2 \theta - 1)$$

$$\psi^{2\pm1} = \sqrt{\frac{15}{8\pi}} \sin \theta \cos \theta (e^{\pm i\phi})$$

$$\psi^{2\pm2} = \sqrt{\frac{15}{32\pi}} \sin^2 \theta (e^{\pm 2i\phi})$$

Equation (1-50) shows that the energy of a rigid rotator is quantized. However, in contrast to the particle-in-a-box and the hydrogen atom problem, it can have zero energy. This is consistent with the Heisenberg uncertainty principle since even if all the momentum is removed from the rotator, the angular position of the rotator is still infinitely variable.

THE LINEAR HARMONIC OSCILLATOR

A harmonic oscillator consists of a particle of mass m acted upon by a force proportional to its displacement x from an equilibrium position. According to Hooke's law, the force F acting on the particle is

$$F \propto -x$$

or

$$F = -kx \tag{1-52}$$

where k is the restoring-force constant. The potential energy V is given by

$$V = -\int F \, dx = \tfrac{1}{2}kx^2 \tag{1-53}$$

Also, from Newton's second law of motion

$$F = m \frac{d^2x}{dt^2} = -kx \tag{1-54}$$

Solving Eq. (1-54) we get

$$x = x_0 \sin \sqrt{\frac{k}{m}} \, t \tag{1-55}$$

The time-independent Schrödinger equation for the harmonic oscillator is

$$\frac{d^2\psi}{dx^2} + \frac{2m}{\hbar^2}(E - \tfrac{1}{2}kx^2)\psi = 0 \tag{1-56}$$

Once again we simply write out the solution of Eq. (1-56):

$$E = (v + \tfrac{1}{2})h\nu \tag{1-57}$$

The wave function for the vth state is given by

$$\psi_v(\xi) = N_v H_v(\xi)e^{-\xi^2/2}$$

where N_v is a normalizing constant given by

$$N_v = \left(\frac{1}{2^v v!}\pi^{-\frac{1}{2}}\right)^{\frac{1}{2}}$$

and ξ is given by $\sqrt{\alpha}t$ where

$$\alpha = \frac{4\pi^2 m\nu}{h}$$

$H_v(\xi)$ is a Hermite polynomial of order v. The number v is the vibrational quantum number; it is zero in the ground state, one in the first excited

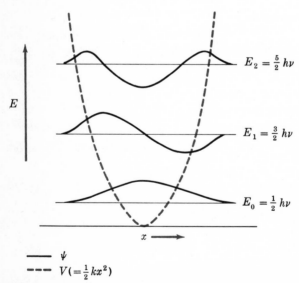

$$E_2 = \tfrac{5}{2}h\nu$$

$$E_1 = \tfrac{3}{2}h\nu$$

$$E_0 = \tfrac{1}{2}h\nu$$

——— ψ

- - - $V(=\tfrac{1}{2}kx^2)$

Fig. 1-9 Plots of ψ and the energy levels of a linear harmonic oscillator.

state, etc.   We have

$$H_0(\xi) = 1$$
$$H_1(\xi) = 2\xi$$
$$H_2(\xi) = 4\xi^2 - 2$$
$$H_3(\xi) = 8\xi^3 - 12\xi$$

The zero-point energy indicated by Eq. (1-57) is again consistent with the Heisenberg uncertainty principle, since if the harmonic oscillator could have zero energy, it would be possible to determine both its momentum and position accurately.   Figure 1-9 shows the energy levels and plots of wave functions for the harmonic oscillator.

1-3   SELECTION RULES

The problems we have just discussed are not merely exercises in quantum mechanics itself; as we shall see in later chapters, with appropriate modifications the results obtained can be used to interpret the changes of rotational, vibrational, and electronic energy levels of molecules.   The common feature here is that the energy of the system is quantized. Hence, given a set of quantized energy levels we might expect that, if appropriate amounts of energy are supplied to the molecule, it would be possible to bring about transition between any two levels.   However, it is found experimentally that the atomic and molecular spectra contain far fewer lines than predicted by the resonance condition, that is, $\Delta E = h\nu$. This must mean that only certain transitions are "allowed" or "preferred." In order to predict the transitions that can occur, we shall first derive the quantum mechanical selection rule, although we shall see that many of the transitions can also be predicted on a more qualitative basis.   We start with the time-dependent treatment as follows: Consider the transition between the m and n stationary states,[†] described by the state functions ψ_m and ψ_n.   Since both ψ_m and ψ_n are solutions of Eq. (1-19), it follows that their linear combination ψ must also be a solution of the same equation.   We write

$$\psi = c_m\psi_m(q,t) + c_n\psi_n(q,t) \tag{1-58}$$

where c_m and c_n are constants.   If the transition from the lower state m to the upper state n is caused by radiation, the hamiltonian $\mathcal{3C}$ in Eq. (1-19) is replaced by

$$\mathcal{3C} \rightarrow \mathcal{3C} + \mathcal{3C}'$$

[†] By stationary state it is meant that the state remains unchanged in the absence of external disturbances.

where $\mathfrak{IC}'$, the perturbation hamiltonian, describes the interaction between the system with the radiation field. We assume that $\mathfrak{IC} \gg \mathfrak{IC}'$. Since transition can now occur between the m and n states, Eq. (1-58) must be rewritten as

$$\psi = c_m(t)\psi_m(q,t) + c_n(t)\psi_n(q,t) \tag{1-59}$$

Equation (1-19) now becomes

$$(\mathfrak{IC} + \mathfrak{IC}')[c_m(t)\psi_m(q,t) + c_n(t)\psi_n(q,t)]$$
$$= i\hbar \frac{\partial[c_m(t)\psi_m(q,t) + c_n(t)\psi_n(q,t)]}{\partial t} \tag{1-60}$$

This gives

$$\mathfrak{IC}[c_m\psi_m(q,t) + c_n\psi_n(q,t)] = i\hbar \left[c_m \frac{\partial\psi_m(q,t)}{\partial t} + c_n \frac{\partial\psi_n(q,t)}{\partial t} \right] \tag{1-61}$$

and

$$\mathfrak{IC}'[c_m(t)\psi_m(q,t) + c_n(t)\psi_n(q,t)]$$
$$= i\hbar \left[\psi_m(q,t) \frac{\partial c_m(t)}{\partial t} + \psi_n(q,t) \frac{\partial c_n(t)}{\partial t} \right] \tag{1-62}$$

We assume that ψ_m and ψ_n are orthonormal, that is

$$\int \psi_n^* \psi_n \, dq = 1 \qquad \int \psi_n^* \psi_m \, dq = 0$$

Multiplying both sides of Eq. (1-62) by $\psi_n^*(q,t)$ and integrating over all space, we get

$$c_m(t) \int \psi_n^*(q,t)\mathfrak{IC}'\psi_m(q,t) \, dq + c_n(t) \int \psi_n^*(q,t)\mathfrak{IC}'\psi_n(q,t) \, dq$$
$$= i\hbar \frac{\partial c_n(t)}{\partial t} \tag{1-63}$$

If the system is initially in the lower state m, we have

$$c_n(t) = 0$$

When the radiation field is applied, the initial rate of transition is given by

$$\frac{dc_n(t)}{dt} = \frac{c_m(t)}{i\hbar} \int \psi_n^*(q,t)\mathfrak{IC}'\psi_m(q,t) \, dq \tag{1-64}$$

For a specific example, let us consider the linear-harmonic-oscillator case. If the particle has a charge e at x and $-e$ at the equilibrium position, then at any moment the electric dipole moment μ is given by

$$\mu = ex \tag{1-65}$$

When an electric field ϵ of the form

$$\epsilon = \epsilon_0(e^{2\pi i\nu t} + e^{-2\pi i\nu t}) \tag{1-66}$$

is present, the hamiltonian $\mathcal{H}'$ representing the electric-field–electric-dipole-moment interaction is given by

$$\mathcal{H}' = \varepsilon \cdot \mathbf{u} \tag{1-67}$$

When the field is just turned on, we still have $c_n(t) = 0$ and $c_m(t) = 1$. Substituting Eq. (1-67) into (1-64) we obtain

$$\frac{dc_n(t)}{dt} = \frac{\epsilon_0}{i\hbar} \left(e^{2\pi i \nu t} + e^{-2\pi i \nu t} \right) \int \psi_n^*(q,t)\mu\psi_m(q,t) \, dq \tag{1-68}$$

Since

$$\psi_m(q,t) = \psi_m(q)e^{-iE_m t/\hbar}$$

and

$$\psi_n(q,t) = \psi_n(q)e^{-iE_n t/\hbar}$$

Eq. (1-68) can be rewritten as a product of the time-dependent part and the space-dependent part, with q replaced by x as follows

$$\frac{dc_n(t)}{dt} = \frac{\epsilon_0}{i\hbar} \left(e^{i(E_n - E_m + h\nu)t/\hbar} + e^{i(E_n - E_m - h\nu)t/\hbar} \right) \mu_{nm} \tag{1-69}$$

where μ_{nm}, the transition dipole moment, is given by

$$\mu_{nm} = \int \psi_n^*(x)\mu\psi_m(x) \, dx$$

Integrating Eq. (1-69) between $t = 0$ and $t = t$, we obtain

$$c_n(t) = \epsilon_0 \left(\frac{1 - e^{i(E_n - E_m + h\nu)t/\hbar}}{E_n - E_m + h\nu} + \frac{1 - e^{i(E_n - E_m - h\nu)t/\hbar}}{E_n - E_m - h\nu} \right) \mu_{nm} \tag{1-70}$$

If the frequency of the electric field radiation ν is such that

$$h\nu = E_n - E_m$$

then only the term containing $E_n - E_m - h\nu$ in Eq. (1-70) is of importance; the term containing $E_n - E_m + h\nu$ is small and can be neglected. Thus

$$c_n(t) = \epsilon_0 \left(\frac{1 - e^{i(E_n - E_m - h\nu)t/\hbar}}{E_n - E_m - h\nu} \right) \mu_{nm}$$

The probability of finding the oscillator in the n state after applying the electric field for time t is given by

$$
\begin{aligned}
c_n^*(t)c_n(t) &= |c_n{}^2| \\
&= \epsilon_0{}^2 \left(\frac{1 - e^{i(E_n - E_m - h\nu)t/\hbar}}{E_n - E_m - h\nu} \right) \left(\frac{1 - e^{-i(E_n - E_m - h\nu)t/\hbar}}{E_n - E_m - h\nu} \right) \mu_{nm}^2 \\
&= 4\epsilon_0{}^2\mu_{nm}{}^2 \frac{\sin^2\left[\pi(E_n - E_m - h\nu)t/h\right]}{(E_n - E_m - h\nu)^2} \tag{1-71}
\end{aligned}
$$

So far we have considered only monochromatic radiation. When there is a range of frequencies present, Eq. (1-71) can be integrated to give

$$|c_n{}^2| = 4\epsilon_0{}^2 \mu_{nm}{}^2 \int_{-\infty}^{+\infty} \frac{\sin^2 [\pi(E_n - E_m - h\nu)t/h]}{(E_n - E_m - h\nu)^2} \, d\nu$$

$$= \frac{\epsilon_0{}^2}{\hbar^2} \mu_{nm}{}^2 t \qquad\qquad\qquad (1\text{-}72)$$

Thus the probability of finding the oscillator in the n state is proportional to the square of the amplitude of the electric field ϵ, the time of irradiation t, and the square of the transition dipole moment. To evaluate μ_{nm} we need to know the explicit wave functions ψ_m and ψ_n as well as $\mathfrak{IC}'$. For the linear harmonic oscillator we can use the wave functions given in Sec. 1-2. However, it is also possible to predict whether certain transitions will be allowed simply by considering the symmetry properties in this case. μ_{nm}, when written as a definite integral, is given by

$$\mu_{nm} = e \int_{-\infty}^{+\infty} \psi_n^*(x) x \psi_m(x) \, dx \qquad\qquad (1\text{-}73)$$

This integral will be zero if the function $\psi_n^*(x) x \psi_m(x)$ is odd and nonzero if the function is even. This follows from the fact that for any function $F(x)$ we write

$$\int_{-\infty}^{+\infty} F(x) \, dx = \int_0^{\infty} F(x) \, dx + \int_{-\infty}^{0} F(x) \, dx$$

$$= \int_0^{\infty} F(x) \, dx + \int_0^{\infty} F(-x) \, dx$$

This integral is nonzero if $F(x)$ is even, that is, $F(x) = F(-x)$ and zero if $F(x)$ is odd, that is, $F(x) = -F(-x)$. Since x itself is an odd function, it follows that $\psi_n(x)$ and $\psi_m(x)$ must be even and odd or odd and even so that μ_{nm} will be nonzero. This establishes the useful results

$$\left.\begin{array}{l} \psi_{\text{even}} \rightarrow \psi_{\text{odd}} \\ \psi_{\text{odd}} \rightarrow \psi_{\text{even}} \end{array}\right\} \quad \text{allowed}$$

$$\left.\begin{array}{l} \psi_{\text{even}} \rightarrow \psi_{\text{even}} \\ \psi_{\text{odd}} \rightarrow \psi_{\text{odd}} \end{array}\right\} \quad \text{forbidden}$$

For the linear-harmonic-oscillator case it turns out that the transitions do obey the above general rule but they are further restricted to adjacent levels only, that is, $m = n - 1$.

So far we have considered only the interaction between the electric field component of the radiation and the electric dipole moment. As we shall see in later chapters there are also electric quadrupole and magnetic dipole interactions. The magnitude of these interactions decreases in the order

Electric dipole interaction $\gg$ magnetic dipole interaction $>$ electric quadrupole interaction

1-4 SPECTROSCOPIC TRANSITIONS

Thus far we have considered only the absorption case. For the reverse case, that is, emission, we write

$$\mu_{mn} = \int \psi_m^*(x)\mu\psi_n(x)\ dx \tag{1-74}$$

(where $\mu_{nm} = \mu_{mn}$). Absorption and emission that take place in the presence of an external field are called the *induced* absorption and emission. In addition, the system in state n may spontaneously lose an amount of energy E given by

$$E = E_n - E_m = h\nu$$

We shall now derive a relationship, according to Einstein, for the transition probabilities of induced absorption and emission and spontaneous emission.

According to the Boltzmann distribution law, the ratio of the population in states n and m, N_n/N_m, is given by

$$\frac{N_n}{N_m} = e^{-(E_n-E_m)/kT} \tag{1-75}$$

where k is the Boltzmann constant. We define the transition probability per unit time for the induced absorption, emission, and spontaneous emission to be B_{mn}, B_{nm}, and A_{nm}. The number of induced emissions per second is proportional to $N_n B_{nm}$ and is given by $N_n B_{nm}\rho(\nu)$, where $\rho(\nu)$ is the density of the radiation of frequency ν; the number of induced absorptions per second is given by $N_m B_{mn}\rho(\nu)$. At steady state, the rates of absorption and emission are equal; hence,

$$N_m B_{mn}\rho(\nu) = N_n B_{nm}\rho(\nu) + N_n A_{nm} \tag{1-76}$$

Since $B_{mn} = B_{nm}$, we write

$$\begin{aligned}
\rho(\nu) &= \frac{N_n A_{nm}}{B_{nm}(N_m - N_n)} \\
&= \frac{A_{nm}}{B_{nm}} \frac{1}{e^{-(E_m-E_n)/kT} - 1}
\end{aligned} \tag{1-77}$$

According to Planck's radiation law

$$\rho(\nu) = \frac{8\pi h\nu^3}{c^3} \frac{1}{e^{h\nu/kT} - 1} \tag{1-78}$$

Substituting Eq. (1-78) into (1-77), we obtain

$$\frac{8\pi h\nu^3}{c^3}\frac{1}{e^{h\nu/kT}-1} = \frac{A_{nm}}{B_{nm}}\frac{1}{e^{-(E_m-E_n)/kT}-1}$$

Since $\nu = \nu_{nm}$, we have

$$\frac{A_{nm}}{B_{nm}} = \frac{8\pi h\nu_{nm}^3}{c^3} \tag{1-79}$$

In general, therefore, the intensity of the absorption lines is proportional to the population in the lower state. Furthermore, when a sample is irradiated there will also be induced and spontaneous emission and we might expect that this competing effect will diminish the intensity of the absorption lines. However, since only the induced emission is coherent with the incident radiation, that is, it travels in the same direction of the radiation field, the *net* absorption intensity is given by

$$N_m B_{mn}\rho(\nu) - N_n B_{nm}\rho(\nu) = B_{mn}\rho(\nu)(N_m - N_n) \tag{1-80}$$

Thus the intensity for a given absorption depends on the difference in the population in states m and n if $\rho(\nu)$ is kept constant.

1-5 ABSORPTION PHOTOMETRY

The equation relating the amount of absorption to the nature of the absorbing species is known as the Beer-Lambert, or Beer's, law

$$\log\frac{I_0}{I} = \epsilon bc \tag{1-81}$$

where I_0 and I are the intensity of the incident and transmitted monochromatic light, b is the light path length in centimeters, c is the molarity, and ϵ is the molar absorbtivity. This equation applies to all types of absorption spectroscopy and holds only for relatively low concentrations ($\lesssim 10^{-2}\ M$). Equation (1-81) is often written as

$$\log\frac{I_0}{I} = A$$

where A is absorbance. For a given b and ϵ, A is directly proportional to the concentration of the absorbing species and therefore Eq. (1-81) is of great importance in quantitative analysis.

1-6 LINEWIDTH AND RESOLUTION

Having obtained the equations for the transition probability (Sec. 1-4), we now turn to the physical appearance of the spectral lines. According to Fig. 1-2 these lines would have no width at all since the energy levels

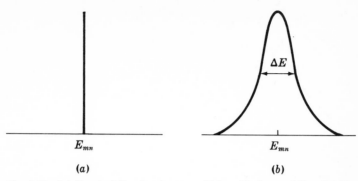

Fig. 1-10   (*a*) Spectral line having no width.   (*b*) Spectral line having width ΔE at half height.

are exactly defined.   However, because of the spontaneous emission, any transition line in reality has a width which is determined by the finite lifetime of the upper state.   This is most clearly shown by the Heisenberg uncertainty principle $\Delta E \; \Delta t \sim \hbar$.   Thus we can no longer represent the upper level by a sharp line (which represents only one value of E); the resulting transition will have an uncertainty in energy ΔE, which is the linewidth† (Fig. 1-10).

In addition to the "lifetime" broadening just discussed, spectral lines can also be broadened by the Doppler effect.   This arises when the molecule being measured has a velocity v relative to the observing instrument.   If the molecule is moving away from the instrument with velocity $-v$, the observed frequency of radiation ν' (by the molecule) is given by

$$\nu' = \nu \left(1 - \frac{v}{c}\right) \tag{1-82}$$

where ν is the frequency of the radiation field and c the velocity of light. Rearranging Eq. (1-82) we obtain

$$\frac{\nu - \nu'}{\nu} = \frac{\Delta \nu}{\nu} = \frac{v}{c} \tag{1-83}$$

On the other hand, if the molecule is moving toward the instrument with velocity v, the observed frequency of radiation ν' (by the molecule) is given by

$$\nu' = \nu \left(1 + \frac{v}{c}\right)$$

or

$$\frac{\nu - \nu'}{\nu} = \frac{\Delta \nu}{\nu} = -\frac{v}{c} \tag{1-84}$$

† We define here linewidth to be the width at half height of the absorption line.

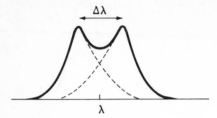

Fig. 1-11 Diagram showing the resolving power of a spectrometer.

Therefore, depending on the direction of motion, the observed frequency is shifted either toward the lower or higher frequency. For an assembly of molecules moving with different velocities relative to the instrument, line broadening will therefore result.

One of the main causes of line broadening is due to the collision among atoms, molecules, and ions. If the time of collision between any two species X and Y is long compared with the mean time between collisions, then the energy of the absorbing or emitting species, say, X, will be perturbed by the electric and magnetic fields due to Y. Consequently the spectral lines of X will be broadened by this disturbance. The proper treatment of this phenomenon can be very complex since it requires the detailed knowledge of intermolecular forces. In many cases, however, this broadening can be related to the pressure of the gas.

Finally, rate processes such as dissociation, rotation, and electron and proton transfer reactions can also cause line broadening. However, we shall not discuss these until later in the appropriate chapters.

Ideally, the analysis of any spectrum requires that the lines be well separated or resolved. This is not always possible although the advantages of a high-resolution spectrum are obvious. In addition to the various mechanisms that can cause line broadening, lines can also be "broadened" by the measuring instrument itself. We define *resolving power* of an instrument as a measure of its ability to distinguish lines that overlap. It is usually expressed as the ratio of the observed wavelength (frequency) to the smallest difference between two wavelengths (frequencies) that can be measured, that is, $\lambda/\Delta\lambda(\nu/\Delta\nu)$ (Fig. 1-11).

1-7 SIGNAL–TO–NOISE RATIO

A recorded spectrum, because of the manner in which the signals are detected, always contains random fluctuations of electronic signals called *noise*. The sensitivity of detecting any real signal depends on how easily we can distinguish the signals from the noise, or the signal-to-noise ratio.† For very weak signals one can employ the computer averaging technique (CAT). Essentially, the signals over a certain region are repeatedly recorded and added up. Since the noise is random, this process will remove much of the noise level and finally strong signals will be obtained.

1-8 UNITS

For any branch of spectroscopy the position of the spectral lines and the separation between them can be measured in terms of frequency, wavelength, or wave number. The wave number $\bar{\nu}$ is defined by

$$\bar{\nu} = \frac{1}{\lambda} = \frac{\nu}{c}$$

It is a more useful quantity than wavelength since it is directly proportional to frequency and can therefore be taken as a measure of the energy. A summary of the units commonly employed is given below:

1. *Frequency* ν
   Cycles per second [cps, c/s, or hertz (Hz)]
   Kilocycles per second (kcps or kHz)
   Megacycles per second (Mcps or MHz)

   $1 \text{ MHz} = 10^3 \text{ kHz} = 10^6 \text{ Hz}$

2. *Wavelength* λ
   Meter (m)
   Centimeter (cm)
   Millimeter (mm)
   Micrometer (μm)
   Nanometer (nm)
   Angstrom (Å)

   $1 \ \mu\text{m} = 10^{-4} \text{ cm} = 10^{-3} \text{ mm}$
   $1 \text{ nm} = 10^{-3} \ \mu\text{m}$
   $1 \text{ Å} = 10^{-8} \text{ cm} = 0.1 \text{ nm}$

3. *Wave number* $\bar{\nu}$
   Number of waves per centimeter (cm^{-1})
   Kayser (K)
   Kilokayser (kK)

   $1 \text{ kK} = 1000 \text{ K} = 1000 \text{ cm}^{-1}$

† The signal measured is usually that of a dc voltage output, whereas the noise is invariably due to an ac source. Thus an ac-to-dc conversion factor is needed in measuring the signal-to-noise ratio.

Table 1-2   Summary of the various branches of spectroscopy

Branch	Phenomenon	Frequency, Hz	Wavelength
Radio frequency	Nuclear magnetic resonance, nuclear quadrupole resonance	10^6 to 10^8	300 to 3 m
Microwave frequency	Electron spin resonance, molecular rotation	10^{10} to 10^{12}	30 to 0.3 m
Infrared	Molecular rotation, molecular vibration	10^{12} to 3×10^{14}	300 to 1 μm
Visible and ultraviolet	Electronic transition (outer electrons)	3×10^{14} to 10^{16}	1 μm to 300 Å
X-rays	Electronic transitions (inner electrons)	3×10^{16} to 10^{19}	100 to 0.3 Å
γ-rays	Nuclear transitions	10^{19} to 10^{22}	0.3 to 0.003 Å

1-9 REGIONS OF SPECTRUM

If we assume that the various types of energy interaction are independent of one another, we can express the total energy E_T of the molecule as

$$E_T = E_{\text{rot}} + E_{\text{vib}} + E_e + \cdots \tag{1-85}$$

This means that the total molecular wave function ψ_T can be written as a product of the individual wave functions

$$\psi_T = \psi_{\text{rot}}\psi_{\text{vib}}\psi_e \cdots \tag{1-86}$$

As we shall see this separation is not exactly true but is a fairly good starting point.

In carrying out any spectroscopic measurement we are usually interested in only one type of energy change. Table 1-2 gives a summary of the energy range of the various branches of spectroscopy of interest to biochemists and chemists.

1-10 CONCLUSION

It is appropriate to point out here that any experimental spectrum (absorption or emission) is actually a superposition of a very large number of spectra of the individual molecules because (1) it is not possible to detect the energy absorbed or emitted by a single molecule, and (2) the interaction between a photon of electromagnetic radiation and a molecule can give rise to only one transition and hence one line. Thus for a spectrum consisting of more than one line (as is usually the case), we are seeing the statistical sum of all the spectra, each of which consists of only one line due to one specific interaction.

PROBLEMS

1-1. Calculate the wavelength associated with (a) an electron moving at 6.6×10^9 cm sec^{-1} and (b) a 10-g ball moving at the same speed.

1-2. Calculate the uncertainty in momentum and velocity of an electron in a one-dimensional box of length 10 Å.

1-3. In the hydrogen atom and hydrogenlike ions the electronic energy levels are only dependent on the principal quantum n, whereas in many-electron atoms the energy levels are dependent on both n and the azimuthal quantum number l. Explain this difference.

1-4. Calculate the Doppler shift in frequency and wavelength for an assembly of molecules moving within the range of velocity of $\pm 1 \times 10^4$ cm sec^{-1}

1-5. Consider the following particle-in-a-box problem: If the particle has a charge of opposite sign distributed symmetrically about the center of the box:
  (a) What is the dipole moment of the system?
  (b) What is the electric transition moment system?
  (c) Deduce the general selection rules for the electric dipole transitions.

SUGGESTIONS FOR FURTHER READING

Introductory

Anderson, J. M.: "Mathematics for Quantum Chemistry," W. A. Benjamin, Inc., New York, 1966.

Hanna, M. W.: "Quantum Mechanics in Chemistry," W. A. Benjamin, Inc., New York, 1966.

Hochstrasser, R. M.: "Behavior of Electrons in Atoms," W. A. Benjamin, Inc., New York, 1965.

Strauss, H. L.: "Quantum Mechanics," Prentice-Hall, Inc., Englewood Cliffs, N.J., 1968.

Intermediate-advanced

Anderson, J. M.: "Introduction to Quantum Chemistry," W. A. Benjamin, Inc., New York, 1969.

Born, M.: "Atomic Physics," Hafner Publishing Company, Inc., New York, 1966.

Eyring, H., J. Walter, and G. E. Kimball: "Quantum Chemistry," John Wiley & Sons, Inc., New York, 1967.

Pauling, L., and E. B. Wilson, Jr.: "Introduction to Quantum Mechanics," McGraw-Hill Book Company, New York, 1935.

Pilar, F. L.: "Elementary Quantum Chemistry," McGraw-Hill Book Company, New York, 1968.

Secs. 1-1 to 1-5

See Anderson; Born; Eyring, Walter, and Kimball; Pauling and Wilson; and Pilar.

Sec. 1-6

Kuhn, H. G.: "Atomic Spectra," p. 381, Academic Press, Inc., New York, 1962.
Townes, C. H., and Schawlow, A. L.: "Microwave Spectroscopy," p. 336, McGraw-Hill Book Company, New York, 1955.

Experimental aspects of spectroscopy

Bair, E. J.: "Introduction to Chemical Instrumentation," McGraw-Hill Book Company, New York, 1962.
Harrison, G. R., R. C. Lord, and J. R. Loofbourow: "Practical Spectroscopy," Prentice-Hall, Inc., Englewood Cliffs, N.J., 1965.

2
Molecular Symmetry and Group Theory

The only chemical system for which the Schrödinger equation can be solved exactly is the hydrogen atom. In general, the quantum mechanical treatment of atoms and molecules can be very complex even when the approximate methods are employed. Fortunately from the symmetry properties of molecules we can obtain very useful information regarding the wave functions and energies without actually solving the Schrödinger equation. Furthermore, from the symmetry of the wave functions we can readily predict the probability of spectroscopic transitions. That all these simplifications are possible is because the use of the symmetry properties can be represented by a special mathematical technique called *group theory*. We shall give a general introduction to group theory and its application to chemistry in this chapter.

Consider the following operation on the fluorine molecule

$$F\!-\!\!\!+\!\!\!-\!F \xrightarrow{180°} F\!-\!F$$

**Table 2-1   Summary of symmetry elements and
symmetry operations**

Symmetry element	Symmetry operation
Rotation axis (proper)	Rotation
Center of symmetry	Inversion
Plane	Reflection
Rotation axis (improper)	Rotation followed by reflection through a plane perpendicular to the rotation axis

If the molecule is rotated by 180° about its center it reaches a new configuration that is indistinguishable from the old one.   Such an operation is called a *symmetry operation*.   That it is possible to carry out a symmetry operation on a molecule means the molecule must possess a *symmetry element*.   A symmetry element is a geometrical entity such as a point, a plane, or a line with respect to which a symmetry operation is performed. Table 2-1 summarizes the various symmetry elements and the corresponding symmetry operations.   We shall now see some examples of each type:

Rotation axis (proper) (C_n) A molecule possesses a proper rotation axis of order n if rotation about the axis by $2\pi/n$ leaves the molecule in an indistinguishable configuration from the original one.   An example of this is the rotation of the BCl_3 molecule.

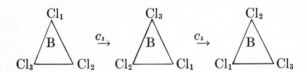

Center of symmetry (i) If the coordinates (x,y,z) of every atom in a molecule are changed into $(-x,-y,-z)$ and the molecule is left in an indistinguishable configuration from the original one, then the point of origin, that is $(0,0,0)$, is the center of symmetry of the molecule.   For example, *trans*-1,2-dichloroethylene possesses a center of symmetry as shown

$$
\begin{array}{ccc}
\text{Cl} & & \text{H} \\
\diagdown & & \diagup \\
& \text{C}\!\!=\!\!\text{C} & \\
\diagup & & \diagdown \\
\text{H} & & \text{Cl}
\end{array}
$$

C_3 C_3

Plane of symmetry (σ)   A molecule possesses a plane of symmetry if the reflection through the plane leaves the molecule unchanged. For example, BCl_3 possesses two planes of symmetry, σ_v and σ_h; the subscript v indicates that the plane contains the highest-order rotation axis, that is, the C_3 axis; the subscript h indicates that the plane is perpendicular to the C_3 axis.

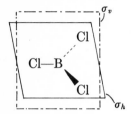

Rotation axis (improper) (S_n)   A molecule possesses an improper rotation axis of order n if rotation about this axis by $2\pi/n$ followed by reflection in a plane perpendicular to the axis leaves the molecule in an indistinguishable position.   An example of this is the allene molecule:

$$\overset{H}{\underset{H}{\diagup}}C=C=C\overset{H}{\underset{H}{\diagdown}} \quad \xrightarrow{C_4} \quad \overset{H}{\underset{H}{\diagup}}C=C=C\overset{H}{\underset{H}{\diagdown}} \quad \xrightarrow{\sigma_v} \quad \overset{H}{\underset{H}{\diagup}}C=C=C\overset{H}{\underset{H}{\diagdown}}$$

Finally we have the identity element E which does not do anything to the molecule and is therefore possessed by all molecules.

Consider now all the symmetry elements of a particular molecule, say, water:

$$\underset{x}{\overset{z}{}}\quad\text{O} \longrightarrow y,\quad \text{H} \quad \text{H}$$

We have E, $C_2{}^z$, $\sigma_v{}^{xz}$, and $\sigma_v{}^{yz}$. It can be readily shown that the product† of any two symmetry elements always gives rise to one of the four elements. For example, $C_2{}^z\sigma_v{}^{xz} = \sigma_v{}^{yz}$. In this way we can construct a multiplication table as shown in Table 2-2.

† Product here means the successive application of the two operations.

Table 2-2   Multiplication table of the symmetry elements of H_2O

	E	$C_2{}^z$	$\sigma_v{}^{xz}$	$\sigma_v{}^{yz}$
E	E	$C_2{}^z$	$\sigma_v{}^{xz}$	$\sigma_v{}^{yz}$
$C_2{}^z$	$C_2{}^z$	E	$\sigma_v{}^{yz}$	$\sigma_v{}^{xz}$
$\sigma_v{}^{xz}$	$\sigma_v{}^{xz}$	$\sigma_v{}^{yz}$	E	$C_2{}^z$
$\sigma_v{}^{yz}$	$\sigma_v{}^{yz}$	$\sigma_v{}^{xz}$	$C_2{}^z$	E

We see that the complete set of operations actually forms a mathematical group. For the elements A, B, C, . . . to form a mathematical group G, the following conditions must hold:

1. There must be a rule of combination such that any elements A and B can combine to give an element C:

   $$AB = C$$

   This rule states that the application of B followed by A is equivalent to the application of C. If $AB = BA$ the elements A and B are said to commute.

2. If the elements A and B are in the same group G and $AB = C$, then C must also be in the same group.

3. The group must contain an element E such that

   $$EA = AE = A \qquad \text{etc.}$$

4. Every element A in the group must have an inverse A^{-1} also in the group such that

   $$AA^{-1} = A^{-1}A = E$$

5. The combination law is associative

   $$A(BC) = (AB)C$$

As a simple example of such a mathematical group, we choose the set of rational numbers between zero and infinity. This set of numbers obviously satisfies the requirements stated above. Thus:

1. $4 \times 7 = 7 \times 4 = 28$

The numbers 4 and 7 are in the same group.

2. The number 28 is also in the same group.

3. $1 \times 6 = 6 \times 1 = 6$

The number 6 is in the same group and the identity element E is 1 in this case.

4. $3 \times \frac{1}{3} = \frac{1}{3} \times 3 = 1$

5. $2 \times (4 \times 6) = (2 \times 4) \times 6$

Before we proceed with the group theory it is appropriate to define here a few terms to be used later.

1. *Order of a group*   This is the number of elements in the group.   In the numerical example we just saw the order is infinity. For the group shown in Table 2-2 the order is only four.
2. *Subgroup*   This is a group within a group.   If a certain selection of the elements of a given group satisfies the definition of a group, then this selection of elements forms a subgroup.   Any subgroup must necessarily include the identity element E.
3. *Similarity transform and conjugate elements*   If A and C are two elements of a group and

$$C^{-1}AC = B$$

where B is also in the same group, then we say that B is the *similarity transform* of A by C and that A and B are *conjugate* to each other.
    It follows that every element is conjugate with itself.   That is

$$E^{-1}AE = A$$

where E is the identity element.   Only if A and C commute can we write

$$C^{-1}AC = A$$

This can be proven as follows.   We multiply both sides by A^{-1}:

LHS: $A^{-1}C^{-1}AC = (CA)^{-1}(AC) = E$
RHS: $A^{-1}A = E$

4. *Class*   A set of elements in a group which are conjugate to one another is said to form a class of the group.

POINT GROUPS

As we saw, the H_2O molecule has four symmetry elements, E, $C_2{}^z$, $\sigma_v{}^{xz}$, and $\sigma_v{}^{yz}$.   If we now consider the NH_3 molecule we will find a different set of symmetry elements.   As it turns out, for the very large number of molecules known to exist there are only relatively few different combinations of the symmetry elements.   In fact, there are only 32 combinations or *point groups* that need concern us in chemistry.   Table 2-3 lists the point groups which are of special interest to spectroscopy.

Table 2-3 Point groups of special interest to spectroscopy

Point group	Symmetry elements	Examples
C_1	E	CHFClBr
C_2	E, C_2	H_2O_2
C_3	E, C_3	C_2H_6
C_{1v}	E, σ_v	NOCl
C_{2v}	$E, C_2, 2\sigma_v$	H_2O
C_{3v}	$E, C_3, 3\sigma_v$	NH_3
$C_{\infty v}$	$E, C_\infty, \infty\sigma_v$	HCl
C_{2h}	E, C_2, σ_h, i	trans-CHF=CHF
D_{2h}	$E, 3C_2, 3\sigma, i$	CH_2=CH_2
D_{3h}	$E, C_3, 3C_2$ ($\perp$ to C_3), $3\sigma_v, \sigma_h$	BCl_3
D_{4h}	$E, C_4, 4C_2$ ($\perp$ to C_4), $4\sigma_v, 4\sigma_h, C_2, S_4$ (coincident with C_4), i	$PtCl_4^{--}$
D_{6h}	$E, C_6, 6C_2$ ($\perp$ to C_6), $6\sigma_v, \sigma_h, C_2, C_3,$ and S_6 (all coincident with C_6), i	C_6H_6
$D_{\infty h}$	E, C_∞, C_2 ($\perp$ to C_∞), $\infty\sigma_v, \sigma_h, i$	F_2
O_h	$E, 3C_4, 4C_3, 3S_4$ and $3C_2$ (both coincident with the C_4 axes), $6C_2, 9\sigma, 4S_6$ (coincident with C_3), i	SF_6
T_d	$E, 3C_2, 4C_3, 6\sigma, 3S_4$ (coincident with C_2)	CH_4

REPRESENTATION OF GROUPS

If we assign each element in the C_{2v} point group (Table 2-2) with a 2×2 matrix† such that

$$E: \begin{bmatrix} 1 & 0 \\ 0 & 1 \end{bmatrix} \quad C_2^z: \begin{bmatrix} -1 & 0 \\ 0 & -1 \end{bmatrix}$$

$$\sigma^{xz}: \begin{bmatrix} 1 & 0 \\ 0 & -1 \end{bmatrix} \quad \sigma^{yz}: \begin{bmatrix} -1 & 0 \\ 0 & 1 \end{bmatrix}$$

we can quickly show that these matrices combine in a manner similar to that shown in Table 2-2; for example:

$$\sigma^{xz}\sigma^{yz} = C_2^z$$

and from matrix multiplication

$$\begin{bmatrix} 1 & 0 \\ 0 & -1 \end{bmatrix} \begin{bmatrix} -1 & 0 \\ 0 & 1 \end{bmatrix} = \begin{bmatrix} -1 & 0 \\ 0 & -1 \end{bmatrix}$$

Thus we say that a representation of a group consists of a set of matrices each of which corresponds to a symmetry element in the group and that these matrices combine in the same way as the symmetry elements do.

† For definition and properties of matrices, see Appendix 1.

However, although the set of matrices shown above do combine in the same way as the symmetry elements, they are certainly not the unique set. There must be many other sets of matrices which combine in the same way and can therefore also be used to represent the C_{2v} group. Our task is to find the most fundamental set of matrices from which other representations are derived. Thus it is very likely that any representation we think of can usually be reduced to the fundamental one by some transformation, and we can now speak of the *reducible* and *irreducible* representations. Symbolically we can illustrate the reducible and irreducible representations as follows: Let A, B, C, . . . be the matrices which form the representation of a group. If there is a matrix X of the same dimension such that

$$X^{-1}AX = A'$$
$$X^{-1}BX = B'$$
. . . .

then the new matrices A', B', . . . also form a representation of the group. If X is the proper transformation matrix, we have

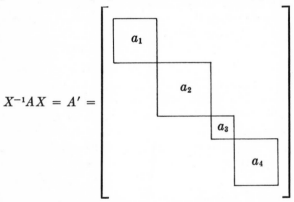

$$X^{-1}AX = A' =$$

The new matrix A' is now in the "blocked-out" form, that is, it consists of smaller matrices along the diagonal with zeros everywhere else. Similarly

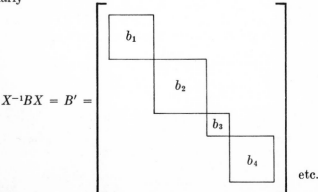

$$X^{-1}BX = B' =$$

etc.

Table 2-4   Irreducible representations
for the C_{2v} point group

	E	$C_2{}^z$	$\sigma_v{}^{xz}$	$\sigma_v{}^{yz}$
Γ_1	1	1	1	1
Γ_2	1	1	-1	-1
Γ_3	1	-1	1	-1
Γ_4	1	-1	-1	1

where a_1, b_1, c_1, . . . , a_2, b_2, c_2, . . . have the same dimensions. These individual matrices multiply in the same way as the matrices A, B, C, . . . or matrices A', B', C', ... Hence,

$$a_1 b_1 = d_1$$
$$a_2 b_2 = d_2$$
$$\cdot \cdot \cdot \cdot \cdot \cdot$$

Thus matrices a_1, b_1, c_1, . . . also form a representation of the group. If at this point it is not possible to find another matrix of the same dimension to carry out further similarity transformations as shown above, we say that a_1, b_1, c_1, . . . form an irreducible representation. Obviously matrices A, B, C, . . . or A', B', C', . . . form a reducible representation.

If we again consider the C_{2v} point group (Table 2-3) and designate the representations as Γ's, we find that there are four irreducible representations, Γ_1, Γ_2, Γ_3, and Γ_4, which satisfy the multiplications (Table 2-4). Each number in Table 2-4 is simply the trace, or character, of a matrix in an irreducible representation of C_{2v}. It is instructive to think of the components of an irreducible representation as orthogonal vectors in an h-dimensional space. Mathematically this is expressed as[1]†

$$\sum_R \Gamma_i(R)_{mn}\Gamma_j(R)_{mn} = \delta_{ij}\frac{h}{l_i} \tag{2-1}$$

$$\sum_R \Gamma_i(R)_{mn}\Gamma_i(R)_{m'n'} = \delta_{mm'}\,\delta_{nn'}\frac{h}{l_i} \tag{2-2}$$

where h is the order of the group, R is an operation of the group, l_i is the dimension of the ith representation, m and n are the row and column, and the δ's have the properties

$$\delta_{ij} = 0 \qquad \text{if } i \neq j$$
$$\delta_{ij} = 1 \qquad \text{if } i = j$$

Since the character of the ith representation of operation R, $\chi_i(R)$, is

† Superscript numbers refer to end-of-chapter references. See Suggestions for Further Reading (preceding References) for full citations, if not given in a particular reference.

just the sum of the diagonal elements of the matrix representing R,

$$\chi_i(R) = \sum_m \Gamma_i(R)_{mm} \tag{2-3}$$

it can be shown that

$$\sum_R \chi_i(R)\chi_j(R) = 0 \qquad i \neq j \tag{2-4}$$

$$\sum_R \chi_i(R)\chi_i(R) = h \tag{2-5}$$

Equations (2-4) and (2-5) can be illustrated using Table 2-4. Thus choosing R's as E and $C_2{}^z$ we have

$$1 \times 1 + 1 \times 1 + 1 \times (-1) + 1 \times (-1) = 0$$

and choosing both R's as $C_2{}^z$, we get

$$1 \times 1 + 1 \times 1 + (-1) \times (-1) + (-1) \times (-1) = 4$$

The importance of Eqs. (2-4) and (2-5) lies in the fact that they can be used to *decompose* any reducible representation into its irreducible components. In general any reducible representation Γ can be written as a linear combination of the irreducible ones:

$$\Gamma = n_1\Gamma_1 + n_2\Gamma_2 + n_3\Gamma_3 + \ldots \tag{2-6}$$

where $\Gamma_1, \Gamma_2, \ldots$ are the irreducible representations and $n_1, n_2, \ldots$ are the number of times each irreducible representation occurs in Γ. We note that

$$\chi(R) = \sum_i n_i\chi_i(R) \tag{2-7}$$

where $\chi(R)$ is the character in the reducible representation which corresponds to the Rth operation. Multiplying Eq. (2-7) by $\chi_j(R)$ and summing over R, we obtain

$$\sum_R \chi_j(R)\chi(R) = \sum_R \sum_i n_i\chi_j(R)\chi_i(R) \tag{2-8}$$

All terms except $i = j$ are zero on the right-hand side [see Eqs. (2-4) and (2-5)] so that we have

$$\sum_R \sum_i n_i\chi_j(R)\chi_i(R) = n_j h$$

and Eq. (2-8) becomes

$$\sum_R \chi_j(R)\chi(R) = n_j h$$

Hence

$$n_j = \frac{1}{h} \sum_R \chi_j(R)\chi(R) \tag{2-9}$$

and

$$n_j = \frac{1}{h} \sum_R a_R \chi_j(R) \chi(R) \tag{2-10}$$

where a_R is the number of elements in the class for which a typical operation is R.

Equations (2-9) and (2-10) are extremely useful for it allows us to determine how many times the jth irreducible representation is contained in a reducible representation. This can be illustrated by considering the reducible representations discussed earlier for C_{2v}.

	E	$C_2{}^z$	$\sigma_v{}^{xz}$	$\sigma_v{}^{yz}$
Γ	$\begin{bmatrix} 1 & 0 \\ 0 & 1 \end{bmatrix}$	$\begin{bmatrix} -1 & 0 \\ 0 & -1 \end{bmatrix}$	$\begin{bmatrix} 1 & 0 \\ 0 & -1 \end{bmatrix}$	$\begin{bmatrix} -1 & 0 \\ 0 & 1 \end{bmatrix}$
$\chi(R)$	2	-2	0	0

Our task is to find how many times Γ_1, Γ_2, Γ_3, and Γ_4 occur in Γ. According to Eq. (2-9) and Table 2-4 we write

$n(\Gamma_1) = \frac{1}{4}[1 \times 2 + 1 \times (-2) + 1 \times 0 + 1 \times 0] = 0$

$n(\Gamma_2) = \frac{1}{4}[1 \times 2 + 1 \times (-2) + (-1) \times 0 + (-1) \times 0] = 0$

$n(\Gamma_3) = \frac{1}{4}[1 \times 2 + (-1) \times (-2) + 1 \times 0 + (-1) \times 0] = 1$

$n(\Gamma_4) = \frac{1}{4}[1 \times 2 + (-1) \times (-2) + (-1) \times 0 + 1 \times 0] = 1$

We have

$$\Gamma = \Gamma_3 + \Gamma_4$$

This is such a simple example that the results can be obtained by inspection. In general we will not even bother to write out the matrices for Γ—only the characters are necessary.

CHARACTER TABLES

Table 2-5 shows the character table for the C_{2v} point group.

Table 2-5   Character table for the C_{2v} point group

C_{2v}	E	$C_2{}^z$	$\sigma_v{}^{xz}$	$\sigma_v{}^{yz}$		
A_1	1	1	1	1	z	x^2, y^2, z^2
A_2	1	1	-1	-1	R_z	xy
B_1	1	-1	1	-1	x, R_y	xz
B_2	1	-1	-1	1	y, R_x	yz

In the left-hand column the old symbols Γ_1, Γ_2, . . . are replaced with the so-called Mulliken symbols. A stands for the one-dimensional representation that is *symmetric* with respect to rotation by $2\pi/n$ about the principal C_n rotation axis, whereas B stands for one-dimensional representation that is *antisymmetric* about the same axis. Thus for A representations we have $\chi(C_n) = 1$ and for B representations $\chi(C_n) = -1$. The subscripts 1 and 2 refer to the representations that are symmetric or antisymmetric with respect to a C_2 axis perpendicular to the principal axis. In the present case there is only one axis of rotation and the subscripts 1 and 2 now refer to whether the representations are symmetric or anti-symmetric to the vertical plane of symmetry. Here we have for A_1, $\chi(\sigma_v{}^{xz}) = 1$ and for A_2, $\chi(\sigma_v{}^{xz}) = -1$. The irreducible representations A_1, A_2, B_1, B_2 are sometimes called the *symmetry species* of the point group. Another symbol, E, which is absent in the column under C_{2v} of Table 2-5 but appears in many of other character tables, stands for two-dimensional representation. In the second column we have all the operations of the group collected at the top and the characters of all the irreducible representations below. In the third column we have the symbols x, y, and z (which are sometimes written as T_x, T_y, and T_z) and R_x, R_y, and R_z. The first three terms refer to translation about the x, y, and z axes. To illustrate this let us consider the vector z which is written in the A_1 row.

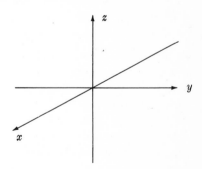

All four applications, that is, E, $C_2{}^z$, $\sigma_v{}^{xz}$, and $\sigma_v{}^{yz}$ on the z vector leave it unchanged. We write

$$E(z) = z$$
$$C_2{}^z(z) = z$$
$$\sigma_v{}^{xz}(z) = z$$
$$\sigma_v{}^{yz}(z) = z$$

This means that the transformation matrices which act on z according to the four operations all have a trace of 1. This is identical to the A irreducible representation of the C_{2v} point group. The appearance of x and y can be similarly explained. We say that z belongs to the A_1

representation or that z transforms as the A_1 representation. The symbols R_x, R_y, and R_z refer to the rotation about the x, y, and z axes. Their appearance in the respective rows in Table 2-5 can be explained by considering the SO_2 molecule which belongs to the C_{2v} point group.

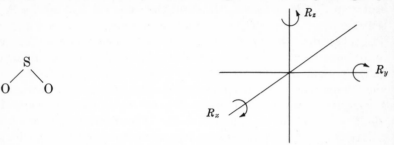

The molecule is in the yz plane. For R_z it is obvious that the operation E leaves it unchanged. The operation $C_2{}^z$ interchanges the two oxygen atoms but does not alter the *sense* of rotation which is anticlockwise. The operation $\sigma_v{}^{xz}$ not only interchanges the two oxygen atoms but also reverses the rotation. The operation $\sigma_v{}^{yz}$ has the same effect on SO_2 as $\sigma_v{}^{xz}$. We have

$$E(R_z) = R_z$$
$$C_2{}^z(R_z) = R_z$$
$$\sigma_v{}^{xz}(R_z) = -R_z$$
$$\sigma_v{}^{yz}(R_z) = -R_z$$

Thus we say that R_z belongs to the A_2 representation. Similarly, R_x and R_y can be shown to belong to the B_2 and B_1 representations, respectively. In the last column we have the symbols xy, xz, yz, x^2, y^2, and z^2. Their transformation properties can be worked out in a similar manner as the examples just discussed although the procedure becomes more involved. Appendix 2 gives all the character tables of interest to spectroscopy and quantum chemistry.

APPLICATIONS OF GROUP THEORY

Spectroscopy   The symmetry property of molecules discussed in this chapter applies only to molecules in their equilibrium, or mean, position, since molecules are continually executing rotational and vibrational motions at all temperatures including absolute zero, at which the vibrational motions still persist. The molecular rotation and vibration energies are quantized, and it is possible to obtain the corresponding spectra (see Chaps. 7 and 8). For a given molecule, the apparently large number of complex vibrations can often be broken down into a relatively small number of specific vibrations or *normal modes* of vibration. The displacement of atoms from their mean positions is defined in terms of the *normal coordinates*, which are functions of angles and distance. In the normal modes of vibration all the atoms move in phase with one another; that is, they pass through their mean positions at the same instant. A very important property of these vibrations is that for nondegenerate normal modes of vibration, the normal coordinates and the vibrational wave functions are either symmetric or antisymmetric with respect to the symmetry operations of the point group of the molecule in its mean position. For degenerate normal modes of vibration the symmetry operations will transform the degenerate set of vibrations into a linear combination of mutually degenerate normal coordinates or wave functions.† In this way we can assign each normal mode of vibration to a symmetry species, or irreducible representation. In Chap. 8 we shall apply the powerful group-theory technique to predict the infrared and Raman activities.

Unlike the vibrational case, the rotational wave functions do not have symmetry properties of interest to spectroscopy.

In electronic spectroscopy we have the electric dipole transitions. The symbols x, y, and z in Table 2-5 can also be taken to represent the components M_x, M_y, and M_z of the dipole-moment vector $\mathbf{M}$. Since these components lie along the x, y, and z axes they must transform in a manner similar to the translation vectors discussed earlier. In Chap. 11 we shall see that the symmetry behavior of the dipole moment can be used to deduce selection rules.

Quantum mechanics Let us consider a quantum mechanical system the Schrödinger wave equation of which is given by

$$\mathcal{H}\psi = E\psi \tag{2-11}$$

To be specific, we consider the helium atom. The hamiltonian is given by

$$\mathcal{H} = \frac{1}{2m_e}\left(\nabla_1{}^2 + \nabla_2{}^2\right) - \frac{2e^2}{r_1} - \frac{2e^2}{r_2} + \frac{e^2}{r_{12}} \tag{2-12}$$

where r_1, r_2, and r_{12} are defined by the following coordinates:

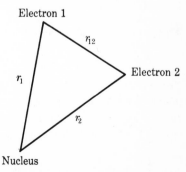

Electron 1

r_{12}

r_1

Electron 2

r_2

Nucleus

† Nondegenerate normal modes of vibration are those that have distinctly different frequencies; degenerate normal modes of vibration are those that have the same frequencies.

The operation of a symmetry element, that is, interchanging the two electrons, gives

$$\mathcal{H} = \frac{1}{2m_e}(\nabla_2{}^2 + \nabla_1{}^2) - \frac{2e^2}{r_2} - \frac{2e^2}{r_1} + \frac{e^2}{r_{21}} \tag{2-13}$$

which obviously leaves the hamiltonian, and hence the energy, unchanged. This operation may, however, affect the wave function ψ. Let this symmetry operation be denoted by P. The hamiltonian in Eq. (2-11) is said to be an *invariant* operator since it commutes with all symmetry operators of the group of the system, that is, the helium atom. Hence

$$\mathcal{H}P - P\mathcal{H} = 0 \tag{2-14}$$

We have

$$P\mathcal{H}\psi = \mathcal{H}P\psi$$

Hence

$$\mathcal{H}(P\psi) = E(P\psi) \tag{2-15}$$

This means that $P\psi$ is itself a solution of Eq. (2-11) and with the same eigenvalue E. If we assume that there is no degeneracy, that is, no two different wave functions have the same energy values, then there is no difference between $P\psi$ and ψ. What is the effect of P on ψ? We write

$$P\psi = C\psi \tag{2-16}$$

where C is a constant. Operating on Eq. (2-16) once more with P we get

$$P^2\psi = \psi = PC\psi = C^2\psi$$

Hence

$$\psi = C^2\psi$$
$$C^2 = 1$$
$$C = \pm 1$$

This gives the result that

$$P\psi = \pm\psi \tag{2-17}$$

Equation (2-17) shows that ψ must be either symmetric or antisymmetric to the exchange of any pair of electrons. Nature has arranged so that the wave function of electrons, protons, and any other particles that have half-integral spins has the property

$$P\psi = -\psi \tag{2-18a}$$

The wave function of particles with integral spins, for example, the deuterium nucleus ($I = 1$), has the property

$$P\psi = \psi \tag{2-18b}$$

Particles described by Eq. (2-18a) are called *fermions*, since they obey the Fermi-Dirac statistics, whereas particles described by Eq. (2-18b) are called *bosons*, since they obey the Bose-Einstein statistics.[2]

The symmetry properties of the wave function just discussed can be shown to lead directly to the Pauli exclusion principle, which is usually stated in a different way: No two electrons in an atom or molecule can have the same four quantum numbers (n, l, m_l, and m_s).

From the above discussion we see that ψ and $P\psi$ both belong to the energy level E. The operation P transforms the eigenfunctions of E into each other. Given this fact we can choose the eigenfunctions of $\mathcal{H}$ with energy E as a basis set for the representation of group G. This choice *always* leads to an irreducible representation of G whether or not E is degenerate. We summarize this in the theorem: If a hamiltonian is invariant under a group G of symmetry operations and G contains all possible symmetry operations of $\mathcal{H}$, then the eigenfunctions belonging to one energy level form a basis for an irreducible representation of G.

Hückel molecular orbital theory Because of the general usefulness of the Hückel molecular orbital (HMO) theory, we shall first give a brief discussion before applying the group-theory method in the calculations.

The HMO theory as applied to the conjugated organic system deals with electrons in the p_π orbital only, and in its most useful form the theory involves some rather drastic approximations. In a qualitative way it is very useful in the study of chemical reactivity, aromaticity, and spectroscopy. Let us consider, as a simple example, the HMO treatment of the butadiene molecule.

$$\overset{1}{C}-\overset{2}{C}-\overset{3}{C}-\overset{4}{C}$$

We write ψ for the molecular orbital, thus

$$\psi = c_1\phi_1 + c_2\phi_2 + c_3\phi_3 + c_4\phi_4 \tag{2-19}$$

where ϕ_1, ϕ_2, . . . are the carbon atomic orbitals containing the p_π electrons and c_1, c_2, . . . are the coefficients. The energy corresponding to the molecular orbital ψ is calculated using the variation principle[3]

$$E = \frac{\int \psi^* \mathcal{H} \psi \, d\tau}{\int \psi^* \psi \, d\tau} \geq E_{\text{true}} \tag{2-20}$$

where E is the energy obtained by using a particular wave function ψ for the molecular orbital and E_{true} is the actual energy of the system.

Equation (2-20) shows that unless we have chosen the correct wave function to start with, the energy calculated will always be greater than the true value. Our task, therefore, is to find the set of coefficients c_i that will yield the lowest energy when put into Eq. (2-20). The condition for this is

$$\frac{\partial E}{\partial c_i} = 0 \qquad\qquad (2\text{-}21)$$

Applying Eq. (2-21) to butadiene in Eq. (2-20) and rearranging, we obtain

$$\frac{\partial E}{\partial c_1} = c_1(\mathfrak{IC}_{11} - ES_{11}) + c_2(\mathfrak{IC}_{12} - ES_{12})$$
$$+ c_3(\mathfrak{IC}_{13} - ES_{13}) + c_4(\mathfrak{IC}_{14} - ES_{14}) = 0$$

$$\frac{\partial E}{\partial c_2} = c_1(\mathfrak{IC}_{21} - ES_{21}) + c_2(\mathfrak{IC}_{22} - ES_{22})$$
$$+ c_3(\mathfrak{IC}_{23} - ES_{23}) + c_4(\mathfrak{IC}_{24} - ES_{24}) = 0 \qquad (2\text{-}22)$$

$$\frac{\partial E}{\partial c_3} = c_1(\mathfrak{IC}_{31} - ES_{31}) + c_2(\mathfrak{IC}_{32} - ES_{32})$$
$$+ c_3(\mathfrak{IC}_{33} - ES_{33}) + c_4(\mathfrak{IC}_{34} - ES_{34}) = 0$$

$$\frac{\partial E}{\partial c_4} = c_1(\mathfrak{IC}_{41} - ES_{41}) + c_2(\mathfrak{IC}_{42} - ES_{42})$$
$$+ c_3(\mathfrak{IC}_{43} - ES_{43}) + c_4(\mathfrak{IC}_{44} - ES_{44}) = 0$$

where

$$\mathfrak{IC}_{ij} = \int \phi_i \mathfrak{IC} \phi_j \, d\tau \qquad S_{ij} = \int \phi_i \phi_j \, d\tau$$

Equation (2-22) represents a set of simultaneous, linear, homogeneous equations in c_i. The condition for such a set of equations to have a nontrivial solution is that the following determinant of the coefficients be zero. Hence

$$\begin{vmatrix} \alpha_1 - ES_{11} & \beta_{12} - ES_{12} & \beta_{13} - ES_{13} & \beta_{14} - ES_{14} \\ \beta_{21} - ES_{21} & \alpha_2 - ES_{22} & \beta_{23} - ES_{23} & \beta_{24} - ES_{24} \\ \beta_{31} - ES_{31} & \beta_{32} - ES_{32} & \alpha_3 - ES_{33} & \beta_{34} - ES_{34} \\ \beta_{41} - ES_{41} & \beta_{42} - ES_{42} & \beta_{43} - ES_{43} & \alpha_4 - ES_{44} \end{vmatrix} = 0 \qquad (2\text{-}23)$$

where

$$\alpha_i = \mathfrak{IC}_{ii}$$
$$\beta_{ij} = \mathfrak{IC}_{ij}$$

This determinant is rather complex and is usually solved by applying the Hückel approximations, which are: (1) α_i, the Coulomb integral, is taken to be the same for all the carbon atoms; (2) β_{ij}, the resonance integral, is assumed to be zero except for adjacent atoms; and (3) S_{ij}, the overlap integral, is taken to be one if $i = j$ and zero if $i \neq j$. Equation (2-23) then becomes

$$
\begin{vmatrix}
\alpha - E & \beta & 0 & 0 \\
\beta & \alpha - E & \beta & 0 \\
0 & \beta & \alpha - E & \beta \\
0 & 0 & \beta & \alpha - E
\end{vmatrix} = 0
\tag{2-24}
$$

Solving Eq. (2-24) we obtain the values for the following four molecular energy levels†

$$
\begin{aligned}
& E_4 = \alpha - 1.618\beta \\
& E_3 = \alpha - 0.618\beta
\end{aligned} \Bigg\} \text{ antibonding molecular orbital}
$$

$$
\begin{aligned}
& E_2 = \alpha + 0.618\beta \\
& E_1 = \alpha + 1.618\beta
\end{aligned} \Bigg\} \text{ bonding molecular orbital}
$$

Substituting the energy values into Eq. (2-22) we get

$$
\begin{aligned}
\psi_1 &= 0.371\phi_1 + 0.600\phi_2 + 0.600\phi_3 + 0.371\phi_4 \\
\psi_2 &= 0.600\phi_1 + 0.371\phi_2 - 0.371\phi_3 - 0.600\phi_4 \\
\psi_3 &= 0.600\phi_1 - 0.371\phi_2 - 0.371\phi_3 + 0.600\phi_4 \\
\psi_4 &= 0.371\phi_1 - 0.600\phi_2 + 0.600\phi_3 - 0.371\phi_4
\end{aligned}
\tag{2-25}
$$

The above procedure can be applied to any unsaturated hydrocarbons and with some modifications (mainly in the choice of α_i and β_{ij}), it can equally well be applied to the heterocyclic systems.

We shall now see how the use of group theory can greatly simplify the calculations. The $2p$ orbitals can be represented by the basis vectors x, y, and z discussed in Table 2-5, and they transform in the same way as the corresponding translation vectors. In the present case (butadiene molecule) we consider only the $2p_\pi$ orbitals, which we assume to be in the z direction. We use these four atomic orbitals (ϕ_1, ϕ_2, ϕ_3, and ϕ_4) as bases for a reducible representation and then decompose them into irreducible representations which correspond to the true eigenfunctions of the molecule. The butadiene molecule belongs to the D_{2h} point group.‡ C_{2h} However, all the necessary symmetry elements are contained in the C_2 character table shown below.

C_2	E	C_2
A	1	1
B	1	-1

† β is a negative quantity.
‡ We assume the molecule to be linear although in reality it can exist in the cis and trans forms.

We now apply the symmetry operations E and C_2 on the four atomic orbitals. If a symmetry operation leaves one of the atomic orbitals unchanged, this function contributes 1 to the character of the operation in the reducible representation; otherwise, it contributes 0. The results are summarized below.

Basis function	E Contribution		C_2 Contribution	
ϕ_1	ϕ_1	1	ϕ_4	0
ϕ_2	ϕ_2	1	ϕ_3	0
ϕ_3	ϕ_3	1	ϕ_2	0
ϕ_4	ϕ_4	1	ϕ_1	0

Thus in the reducible representation Γ we have

$$\chi(E) = 4$$
$$\chi(C_2) = 0$$

The next step is to find out how many times each of the irreducible representations A and B occur in Γ. This can be done by using Eq. (2-9).

$$n_A = \tfrac{1}{2}(1 \times 4 + 1 \times 0) = 2$$
$$n_B = \tfrac{1}{2}(1 \times 4 - 1 \times 0) = 2$$

This gives

$$\Gamma = 2A + 2B$$

which means that A contains two *symmetry orbitals* ψ_1' and ψ_2' and B contains two *symmetry orbitals* ψ_3' and ψ_4'. These symmetry orbitals are obtained by using the so-called basis-generating-machine procedure as follows: We apply each symmetry operation in turn to each nonequivalent atomic orbital and multiply the atomic orbital by the character. Thus

$$\psi_i = \sum_R R\phi_j\chi_i(R)$$

For butadiene we have, for the A irreducible representation,

For ϕ_1: $\psi_1' = E\phi_1\chi(E) + C_2\phi_1\chi(C_2)$
$$= \phi_1 + \phi_4$$
For ϕ_2: $\psi_2' = E\phi_2\chi(E) + C_2\phi_2\chi(C_2)$
$$= \phi_2 + \phi_3$$

It can be readily shown that starting with ϕ_3 and ϕ_4 will not give us independent symmetry orbitals since ϕ_3 is equivalent to ϕ_2 and ϕ_4 to ϕ_1.

Similarly for the B irreducible representation we have

For ϕ_1: $\quad \psi_3' = E\phi_1\chi(E) + C_2\phi_1\chi(C_2)$
$\qquad\qquad = \phi_1 - \phi_4$
For ϕ_2: $\quad \psi_4' = E\phi_2\chi(E) + C_2\phi_2\chi(C_2)$
$\qquad\qquad = \phi_2 - \phi_3$

After normalization, the four symmetry orbitals are

$$\psi_1' = \frac{1}{\sqrt{2}}(\phi_1 + \phi_4) \qquad \psi_3' = \frac{1}{\sqrt{2}}(\phi_1 - \phi_4)$$
$$\psi_2' = \frac{1}{\sqrt{2}}(\phi_2 + \phi_3) \qquad \psi_4' = \frac{1}{\sqrt{2}}(\phi_2 - \phi_3)$$

(2-26)

The energies of the molecule can now be obtained by solving two 2×2 determinants instead of one 4×4 determinant as shown in Eq. (2-24). For A we have

$$\mathcal{H}_{11} = \int \psi_1 \mathcal{H} \psi_1 \, d\tau = \tfrac{1}{2} \int (\phi_1 + \phi_4)\mathcal{H}(\phi_1 + \phi_4) \, d\tau$$
$$= \alpha$$

$$\mathcal{H}_{12} = \mathcal{H}_{21} = \int \psi_1 \mathcal{H} \psi_2 \, d\tau = \tfrac{1}{2} \int (\phi_1 + \phi_4)\mathcal{H}(\phi_2 + \phi_3) \, d\tau$$
$$= \beta$$

$$\mathcal{H}_{22} = \int \psi_2 \mathcal{H} \psi_2 \, d\tau = \tfrac{1}{2} \int (\phi_2 + \phi_3)\mathcal{H}(\phi_2 + \phi_3) \, d\tau$$
$$= \alpha + \beta$$

This gives the determinant

$$\begin{vmatrix} \alpha - E & \beta \\ \beta & \alpha + \beta - E \end{vmatrix} = 0$$

(2-27)

Similarly for B we get

$$\mathcal{H}_{33} = \alpha$$
$$\mathcal{H}_{34} = \mathcal{H}_{43} = \beta$$
$$\mathcal{H}_{44} = \alpha - \beta$$

This gives the determinant

$$\begin{vmatrix} \alpha - E & \beta \\ \beta & \alpha - \beta - E \end{vmatrix} = 0$$

(2-28)

Solution of these two determinants then gives the four energy values as previously obtained from Eq. (2-24). We note that Eq. (2-26) does not correspond to Eq. (2-25). The reason is that instead of the symmetry orbitals themselves we have obtained the linear combination of the symmetry orbitals in Eq. (2-26).   Thus we have

$$\begin{aligned}
\psi_1 &= a_1(\psi_1' + \psi_4') + a_2(\psi_2' + \psi_3') \\
\psi_2 &= a_2(\psi_1' + \psi_4') - a_1(\psi_2' + \psi_3') \\
\psi_3 &= a_3(\psi_1' - \psi_4') + a_4(\psi_2' - \psi_3') \\
\psi_4 &= a_3(\psi_1' - \psi_4') - a_4(\psi_2' - \psi_3')
\end{aligned} \qquad (2\text{-}29)$$

where $a_1, a_2, \ldots$ are the coefficients in the secular equation

$$\begin{aligned}
(\alpha - E)a_1 + \beta a_2 &= 0 \\
\beta a_1 + (\alpha + \beta - E)a_2 &= 0 \\
(\alpha - E)a_3 + \beta a_4 &= 0 \\
\beta a_3 + (\alpha - \beta - E)a_4 &= 0
\end{aligned} \qquad (2\text{-}30)$$

The determinant of the coefficients in Eq. (2-30) is given by Eqs. (2-27) and (2-28).   Thus, substituting the energy values into Eq. (2-30), we get the coefficients $a_1, a_2, \ldots$, which when substituted into Eq. (2-29) give the proper symmetry orbitals.

With the butadiene molecule we have illustrated the basic steps of applying group theory in HMO theory.   The main reason for classifying the molecular orbitals according to the different irreducible representations is that matrix elements of the type $\int \psi_i \mathcal{3C} \psi_j \, d\tau$ are nonzero only if ψ_i and ψ_j belong to the same irreducible representation.   In this way we can always factor the original $n \times n$ determinant into smaller determinants and thereby greatly simplify the mathematics.

PROBLEMS

2-1. Verify all the symmetry elements shown in Table 2-3.

2-2. Classify the following molecules according to their point groups: pyridine, CH_3Cl, ferrocene, CH_3D, $Fe(CN)_6^{--}$, B_2H_6, and $\cdot CH_3$ (planar).

2-3. The biphenyl molecule is planar in the solid phase, but in the vapor phase the two phenyl rings are twisted out of the plane (about 40°).   Also, in some o,o'-substituted biphenyls the two rings are perpendicular to each other due to steric hindrance. Obtain the point group for each case.

2-4. A necessary condition for optical activity is that the molecule must not possess an S_n axis.   Apply this rule to the following molecules: CH_3Cl, CHFIBr, cis-1,2-dibromocyclobutane, and 2,2'-dimethylbiphenyl (the angle of twist is less than 90°).

SUGGESTIONS FOR FURTHER READING

Introductory

Cotton, F. A.: "Chemical Applications of Group Theory," Interscience Publishers, a division of John Wiley & Sons, Inc., New York, 1963.

Jaffe, H. H., and M. Orchin: "Symmetry in Chemistry," John Wiley & Sons, Inc., New York, 1967.

Intermediate-advanced

Hochstrasser, R. M.: "Molecular Aspects of Symmetry," W. A. Benjamin, Inc., New York, 1966.

READING ASSIGNMENTS

Group theory

An Introduction to Molecular Symmetry and Symmetry Point Groups, M. Zeldin, *J. Chem. Educ.*, **43**:17 (1966).

An Introduction to Group Theory for Chemists, J. E. White, *J. Chem. Educ.*, **44**:128 (1967).

Symmetry, Point Groups, and Character Tables, M. Orchin and H. H. Jaffe, *J. Chem. Educ.*, **47**:246, 372, 510 (1970).

Molecular orbital theory

Cotton, p. 117.

"Molecular Orbital Theory for Organic Chemists," A. Streitwieser, Jr.: chaps. 2 and 3, John Wiley & Sons, Inc., New York, 1961.

REFERENCES

1. Cotton, p. 63.
2. D. F. Eggers, N, W. Gregory, G. D. Halsey, and B. S. Rabinovitch, "Physical Chemistry," p. 346, John Wiley & Sons, Inc., New York, 1964.
3. H. L. Strauss, "Quantum Mechanics," p. 105, Prentice-Hall, Inc., Englewood Cliffs, N.J., 1968.

See problem 6-2.

3
Nuclear Magnetic Resonance Spectroscopy

3-1 INTRODUCTION

The fundamental properties of a proton are: (1) mass, (2) charge, and (3) spin or intrinsic angular momentum. The spinning motion of the proton acts like a circular current and so generates a magnetic field. But the magnetic field of a circular current is equivalent to that of a magnetic dipole of moment μ_I given by (see Appendix 3)

$$\mu_I = g_N \frac{e}{2M_p} \hbar \sqrt{I(I+1)} \tag{3-1}$$

where g_N is the nuclear g factor, M_p is the mass of the proton, e is the electronic charge, and $\sqrt{I(I+1)}\ \hbar$ is the length or magnitude of the spin-angular momentum vector.

When a proton is introduced into a uniform magnetic field of strength H, the nuclear magnetic dipole will precess about the axis of the field. This precession occurs because there is a tendency for the magnetic field to turn the nuclear magnetic moment around into the field direction. But this effort is opposed by the rotational inertia due to the revolution of the proton, and the consequent motion is analogous to the precession of a spinning mechanical top under the influence of gravity. The angular precession frequency ω, which is called the Lamor frequency, is given by

$$\omega = \gamma H \tag{3-2}$$

where γ, the gyromagnetic ratio, is defined by

$$\gamma = \frac{\text{magnetic moment}}{\text{angular momentum}} = \frac{\mu_I}{\sqrt{I(I+1)}\,\hbar} \tag{3-3}$$

The energy E due to this magnetic interaction is given by

$$E = -\mu_I H \cos \theta \tag{3-4}$$

where θ is the angle between the axis of the dipole and the field direction. Classically, θ can take any value, so that the energy varies continuously. Quantum mechanics, however, imposes the restriction that the angular momentum is quantized in space so there are only certain allowed values for θ. Figure 3-1 shows the vector representation of the relation between the angular momentum and its components along the axis of quantization, that is, the external magnetic field.

For a proton having spin I of $\frac{1}{2}$ there are only two possible values of θ. The projection of the spin angular momentum vector onto the axis of quantization gives $\frac{1}{2}\hbar$ and $-\frac{1}{2}\hbar$. This can be written as $M_I\hbar$ where M_I is the magnetic spin quantum number which has the values of $\pm\frac{1}{2}$. It can be readily shown that θ in this case is either $35°15'$ or $144°45'$. The same is also true for the magnetic moment vector.

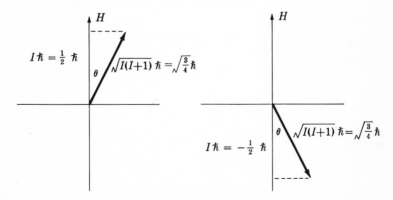

Fig. 3-1 Nuclear spin-angular-momentum vector and its components along the external magnetic field.

If μ_H is the component of μ_I in the direction of the field, we have

$$\cos\theta = \frac{\mu_H}{\mu_I} \tag{3-5}$$

Equation (3-4) now becomes

$$E = -\mathbf{\mu}_H \cdot \mathbf{H} \tag{3-6}$$

where μ_H is given by

$$\mu_H = g_N\beta_N M_I \tag{3-7}$$

where β_N is the nuclear magneton, equal to $(e/2M_pc)\hbar$ (see Appendix 3). Substituting Eq. (3-7) into (3-6) we obtain

$$E = -g_N\beta_N H M_I \tag{3-8}$$

Thus for a spin of $\frac{1}{2}$, $M_I = \pm\frac{1}{2}$, and it has a lower energy level of $-\frac{1}{2}g_N\beta_N H$ which corresponds to the parallel alignment of the magnetic moment to the applied field and a higher energy level of $\frac{1}{2}g_N\beta_N H$ which corresponds to the antiparallel situation.† In the absence of the field these two levels are said to be *degenerate*. This degeneracy is removed when the external field is applied; the splitting of the levels is shown in Fig. 3-2.

† The direction of the magnetic moment is the direction from the south pole of the magnet to the north pole.

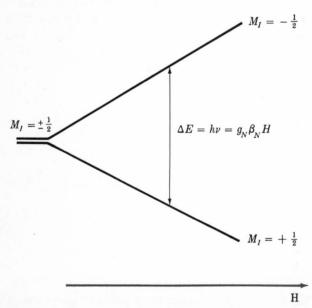

Fig. 3-2 The splitting of the nuclear Zeeman levels in a magnetic field.

If electromagnetic radiation of frequency ν is present which satisfies the resonance condition

$$\Delta E = h\nu = g_N \beta_N H \tag{3-9}$$

transition between these Zeeman levels can occur. This is the simplest case of nuclear magnetic resonance (NMR). For a system containing an assembly of noninteracting protons, the ratio of the proton population in the $M_I = -\frac{1}{2}$ state, $n_{-\frac{1}{2}}$, to that in the $M_I = \frac{1}{2}$ state, $n_{+\frac{1}{2}}$, is given by Boltzmann's expression

$$\frac{n_{-\frac{1}{2}}}{n_{+\frac{1}{2}}} = e^{-g_N \beta_N H / kT} \tag{3-10}$$

The nuclear spin I for any nucleus may be zero, a half-integer, or an integer—its value is determined by the mass number A and atomic number Z of the nucleus. The following rules are useful in determining the value of I:

1. If A is odd and Z is even or odd, I is a half-integer.
2. If A and Z are both even, I is zero.
3. If A is even and Z is odd, I is an integer.

Appendix 4 gives a list of the magnetic moments and spins of all the elements of interest in chemical spectroscopy.

If I is zero, μ_I is zero [Eq. (3-1)], and NMR experiments are not possible. Some examples of nuclei with nonzero spins, for which NMR is applicable, are C^{13}, N^{14}, N^{15}, P^{35}, etc., but we shall concentrate only on the proton magnetic resonance in this chapter.

3-2 EXPERIMENTAL TECHNIQUES

NMR experiments are performed by detecting the amount of energy absorbed; therefore, in order to improve the sensitivity of detection, we want to have as many spins in the lower state as possible. According to Eq. (3-10) this can be achieved by either reducing the temperature or increasing the field strength, or both. There are, of course, practical limitations to these two variables.

The basic features of an NMR spectrometer are: (1) a source of radio-frequency radiation, (2) a receiver coil, (3) a dc magnetic field, and (4) a recorder or an oscilloscope. Figure 3-3 shows the schematic diagram of an NMR spectrometer.

The sample tube is placed between the poles of a powerful magnet. The source of energy, that is, the radio-frequency field, is generated in the coil connected to the r-f oscillator. The detector coil is placed at right angles to both the direction of the magnetic field and the transmitter

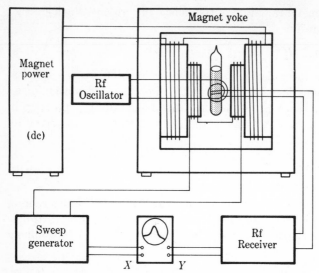

Fig. 3-3 Block diagram showing NMR spectrometer equipped with an electromagnet. (*By permission of J. D. Roberts, "Nuclear Magnetic Resonance," McGraw-Hill Book Company, New York, 1959.*)

coil. The magnet is provided with the sweep coils which are used to vary the field over a range of a few gauss. In a typical experiment, we fix the frequency of the r-f field and vary the magnetic field H until the resonance condition is reached. The nuclear-magnetic-moment transition induces an emf in the detector coil which is then amplified and displayed on the recorder or oscilloscope.

If the magnetic field H is not homogeneous over the sample tube, resonance will occur at apparently different magnetic fields and line broadening will result because of the overlapping of lines. This effect can be minimized by spinning the sample tube with an appropriate frequency so that all the nuclei will experience an average magnetic field. The frequency of spinning should be greater than the rate at which the magnetic field fluctuates.

3-3 THEORY

THE CHEMICAL SHIFT

According to the discussion in Sec. 3-1 we should expect all the protons to absorb energy in the same magnetic field. In molecules, however, this is hardly the case. Consider the low-resolution spectrum of ethanol shown in Fig. 3-4. Three peaks are obtained which have the relative intensities

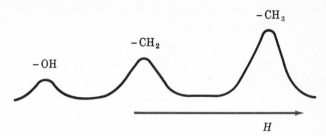

Fig. 3-4   Low-resolution NMR spectrum of ethanol.

$1:2:3$ corresponding to one, two, and three protons.   This implies that the actual, or effective, fields at the nuclei are different from the applied field H_0.   If H is the field at the nucleus we write

$$H = H_0(1 - \sigma) \tag{3-11}$$

where σ is a dimensionless constant called the *screening constant* and is of the order of 10^{-5} for protons.   Figure 3-5 shows the modification of the resonance condition for a proton in going from a bare nucleus to the screened one.

The value of σ depends only on the chemical environment of the nucleus, that is, the electronic configuration around the nucleus.   Consider the following two resonance lines due to nuclei A and B (Fig. 3-6). The difference between the resonance fields or frequencies of A and B is called the chemical shift.   Hence

$$\begin{aligned} H_B - H_A &= H_0(1 - \sigma_B) - H_0(1 - \sigma_A) \\ &= H_0(\sigma_A - \sigma_B) = H_0 \delta_{AB} \end{aligned} \tag{3-12}$$

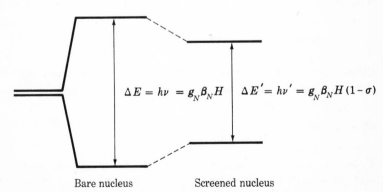

$$\Delta E = h\nu = g_N \beta_N H \qquad \Delta E' = h\nu' = g_N \beta_N H(1-\sigma)$$

Bare nucleus   Screened nucleus

Fig. 3-5   Modification of nuclear Zeeman level splitting as a result of electronic screening.

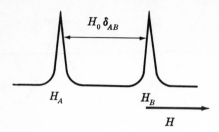

Fig. 3-6 The chemical shift between lines A and B.

and

$$\nu_B - \nu_A = \nu_0(1 - \sigma_B) - \nu_0(1 - \sigma_A)$$
$$= \nu_0(\sigma_A - \sigma_B) = \nu_0 \delta_{AB} \qquad (3\text{-}13)$$

where δ_{AB}, the chemical shift, is given by

$$\delta_{AB} = (\sigma_A - \sigma_B) \qquad (3\text{-}14)$$

δ_{AB} is usually expressed in frequency as

$$\delta_{AB} = \frac{\nu_B - \nu_A}{\nu_0} \times 10^6 \text{ parts per million (ppm)} \qquad (3\text{-}15)$$

Since ν_A and ν_B are almost equal and ν_0, the frequency at which the spectrometer operates (in the megahertz range), a factor of 10^6 is therefore included in Eq. (3-15) so that δ_{AB} can be expressed in convenient numbers.

In practice, chemical shifts of protons in organic compounds are usually given with reference to tetramethylsilane (TMS), $(CH_3)_4Si$. This compound is used as an internal standard because of the following features: It is chemically inert; it contains 12 protons of the same type so that only a small amount of it would be needed for a single, intense line. The chemical shift of a sample with reference to TMS is denoted by

$$\delta_{\text{sample,TMS}} = \frac{\nu_{\text{sample}} - \nu_{\text{TMS}}}{\nu_0} \times 10^6 \text{ ppm} \qquad (3\text{-}16)$$

Since the resonance of TMS occurs at a higher field than most protons, a different chemical shift τ is sometimes employed in order to avoid the use of negative numbers.

$$\tau = 10.00 + \delta_{\text{sample,TMS}} \qquad (3\text{-}17)$$

Unfortunately there does not appear to be an agreed convention for quoting chemical shifts and care must be given when comparing shift measurements in literature. Table 3-1 gives the τ values for protons in some typical organic compounds.

Table 3-1   Some typical values of τ for protons†

Type of protons	τ
TMS	10.0
$(CH_3)_4C$	9.0
CH_3NO_2	5.7
C_6H_6	2.7
$CH_2{=}CH_2$	4.5
$CH{\equiv}CH$	7.8
CH_4	9.8
CH_3I	7.8
CH_3Br	7.3
CH_3Cl	7.0
CH_3F	5.7

† Relative to TMS.

In general, there are three types of contributions to the screening constant

$$\delta = \sigma_{\text{diamagnetic}} + \sigma_{\text{paramagnetic}} + \sigma_{\text{solvent}} \qquad (3\text{-}18)$$

$\sigma_{\text{diamagnetic}}$ is related to the induced circulation of electrons which generates a magnetic field opposed to the applied field H_0. This is a diamagnetic effect, the result of which is the shielding of the nucleus by the electrons from the field. $\sigma_{\text{paramagnetic}}$ is related to the anisotropy in the electron distribution about the atom. In this case, the electron circulation about the nucleus will generate a magnetic field in the *same* direction as the applied field, the result of which is the deshielding of the nucleus; that is, $\sigma_{\text{paramagnetic}}$ is negative. Finally, the chemical shift of a solute molecule in certain solvents will also depend on the nature of the solvent molecules. The effect of the shielding is represented by σ_{solvent}, which can often be eliminated by measuring the chemical shift of the solute molecules over a range of concentrations and extrapolating to zero concentration. For protons in organic molecules only $\sigma_{\text{diamagnetic}}$ is of importance. Theoretical calculations of $\sigma_{\text{diamagnetic}}$ are quite difficult, but qualitatively we can say that it is proportional to the charge densities.

Aromatic protons usually appear at lower fields than aliphatic ones; this is believed to be due to the so-called ring-current effect, as follows: Consider the benzene molecule in a magnetic field H (Fig. 3-7). The circulating current in benzene due to the six π electrons produces a magnetic dipole; however, the induced field at the position of the protons is parallel to the applied field and the result is therefore deshielding. Although such crude models can lead to quite reasonable shift estimates, it is unwise to use experimental shifts as evidence for aromaticity.

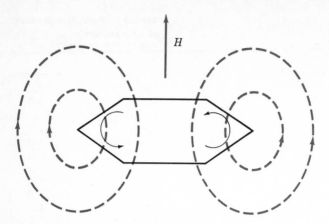

Fig. 3-7 Ring current and magnetic lines of force in benzene.

DIPOLAR INTERACTION

Consider a system containing two protons separated by distance r. The interaction between the two magnetic dipoles can be represented by the hamiltonian $\mathcal{K}_{dip}$, which is given by[2]

$$\mathcal{K}_{dip} = g_N{}^2\beta_N{}^2 \left[\frac{\mathbf{I}_1 \cdot \mathbf{I}_2}{r^3} - \frac{3(\mathbf{I}_1 \cdot \mathbf{r})(\mathbf{I}_2 \cdot \mathbf{r})}{r^5}\right] \tag{3-19}$$

where $\mathbf{I}_1$ and $\mathbf{I}_2$ are the nuclear spin quantum numbers of protons 1 and 2. As a result of this interaction there will now be an additional field at proton 1, H_{loc}, due to proton 2. We have

$$H_{loc} = \pm\frac{3g_N\beta_N}{4}\frac{1 - 3\cos^2\theta}{r^3} \tag{3-20}$$

where θ is the angle between the applied field H and the line joining the nuclei. The $\pm$ sign accounts for the fact that the magnetic dipole of proton 2 can be aligned parallel or antiparallel to proton 1. The same applies to proton 2. Therefore, the effective field at each nucleus, H_{eff}, is given by

$$H_{eff} = H_0 \pm\frac{3g_N\beta_N}{4}\frac{1 - 3\cos^2\theta}{r^3} \tag{3-21}$$

Consequently the absorption spectrum will be a doublet with a separation ΔH of

$$\Delta H = \tfrac{3}{2}g_N\beta_N\frac{1 - 3\cos^2\theta}{r^3} \tag{3-22}$$

An interesting example of the NMR study of solids was shown by Pake[2] who found that the NMR spectrum of the single-crystal gypsum,

$CaSO_4 \cdot 2H_2O$, gave rise to four lines (two doublets). Since the interaction between each pair of equivalent protons will only give rise to a doublet, the conclusion is that the two water molecules in each unit have different orientations.

In nonviscous liquids this type of magnetic interaction vanishes because of the rapid tumbling motion of the molecules. The reason is that when a molecule tumbles in solution, the angle θ is constantly changing, and if the frequency of tumbling is equal to or greater than the energy resulting from the dipolar interaction expressed in frequency, this effect will be averaged to zero.

SPIN–SPIN INTERACTION

When the NMR spectrum of ethanol is recorded under high resolution it is found that each of the three peaks shown in Fig. 3-4 is split into multiplets (Fig. 3-8). The splitting of the lines arises as a result of the nuclear spin-spin interaction as follows: For a proton we have $M_I = \pm\frac{1}{2}$, and if we denote α for $M_I = \frac{1}{2}$ and β for $M_I = -\frac{1}{2}$, then the possible combinations of spin orientation for the $—CH_2$ and $—CH_3$ groups are:

ΣM_I	$—CH_2$ group	Degeneracy	ΣM_I	$—CH_3$ group	Degeneracy
$+1$	$\alpha\alpha$	1	$\frac{3}{2}$	$\alpha\alpha\alpha$	1
0	$\alpha\beta, \beta\alpha$	2	$\frac{1}{2}$	$\alpha\alpha\beta, \alpha\beta\alpha, \beta\alpha\alpha$	3
-1	$\beta\beta$	1	$-\frac{1}{2}$	$\beta\beta\alpha, \beta\alpha\beta, \alpha\beta\beta$	3
			$-\frac{3}{2}$	$\beta\beta\beta$	1

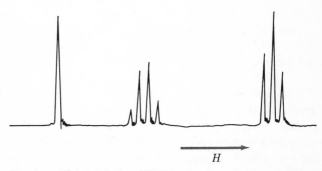

H

Fig. 3-8 High-resolution NMR spectrum of ethanol with a trace of acid added. The spin-spin interaction between hydroxyl proton and the $—CH_2$ group is discussed in Sec. 3-4. (*By permission of J. D. Roberts, "Nuclear Magnetic Resonance," McGraw-Hill Book Company, New York, 1959.*)

The three possible orientations of the methylene protons will cause three different local fields at the methyl protons and therefore split the methyl absorption peak into three lines of relative intensities 1:2:1. Likewise, the methylene peak is split into four lines of relative intensities 1:3:3:1 by the methyl protons. Thus we have

$-CH_2$ $-CH_3$

In general if a line is split by n magnetically equivalent protons (to be defined later), there will be $n + 1$ lines: the relative intensities of the lines are proportional to the coefficients of a binomial expansion of order n, that is, $(1 + x)^n$. As we shall see later, although there is also spin-spin interaction between protons of the same group (methylene or methyl), this effect cannot be observed since the protons are magnetically equivalent.

It is important to note here that the splitting due to this spin-spin interaction, unlike the chemical shift, is independent of the applied field. This is shown by the NMR spectrum of $Br-CH_2-CH_2-CH_2-CN$ recorded at three different fields (Fig. 3-9). The three groups of lines corresponding to three different $-CH_2$ groups spread farther apart at higher fields (chemical shift), but the spacings within each group (spin-spin interaction) remain constant.

This spin-spin interaction is not a direct dipole-dipole interaction, as discussed above, since its magnitude is much too small, and it does not depend on the orientation so it can be measured in liquids.   Instead, this interaction is due to an indirect electron-coupled nuclear-spin interaction. Consider a system containing two nuclei A and B both having I of $\frac{1}{2}$.   In the presence of a magnetic field H along the z direction, the hamiltonian $\mathcal{3C}$ for the various magnetic interactions is given by

$$\mathcal{3C} = \mathcal{3C}_{Zeeman} + \mathcal{3C}_{sc} + \mathcal{3C}_{dip} \tag{3-23}$$

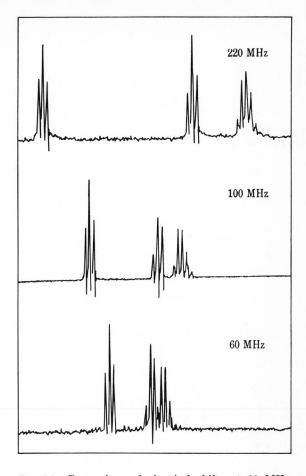

Fig. 3-9 Comparison of chemical shifts at 60 MHz, 100 MHz, and 220 MHz for $BrCH_2CH_2CH_2CN$. [*From J. K. Becconsall and M. C. McIvor, Chem. Brit.*, **5**: *147 (1969). By permission of the authors and the Royal Institute of Chemistry and the Chemical Society.*]

$\mathcal{H}_{Zeeman}$ represents the interaction of the nuclei with the applied field H_0.

$$\mathcal{H}_{Zeeman} = -g_N\beta_N I_z{}^A H_0(1 - \sigma_A) - g_N\beta_N I_z{}^B H_0(1 - \sigma_B) \qquad (3\text{-}24)$$

where $I_z{}^A$ and $I_z{}^B$ are the components of the nuclear-spin quantum numbers in the z direction. $\mathcal{H}_{sc}$, where sc stands for scalar coupling, represents the indirect spin-spin interaction and is given by

$$\mathcal{3C}_{sc} = J\mathbf{I}^A \cdot \mathbf{I}^B \tag{3-25}$$

where J, the coupling constant, is a measure of the extent of the interaction and given in hertz. $\mathcal{3C}_{dip}$ takes the same form as Eq. (3-19). However, if we are dealing with the system in the liquid state, as is usually the case in practice, it averages to zero. It is convenient to work in frequency units so that Eq. (3-23) is rewritten as

$$\mathcal{3C} = -\nu_0(1 - \sigma_A)I_z{}^A - \nu_0(1 - \sigma_B)I_z{}^B + J\mathbf{I}^A \cdot \mathbf{I}^B \tag{3-26}$$

where

$$\mathbf{I}^A \cdot \mathbf{I}^B = I_x{}^A I_x{}^B + I_y{}^A I_y{}^B + I_z{}^A I_z{}^B \tag{3-27}$$

The energy due to the interactions can be obtained by solving the Schrödinger equation

$$\mathcal{3C}\psi = E\psi \tag{3-28}$$

where ψ is the exact wave function and can be written as a linear combination of the basis spin functions ϕ_i

$$\psi = \sum_i c_i\phi_i \tag{3-29}$$

For a two-nuclei system having $I = \frac{1}{2}$ there are four possible combinations of spins given by

$$\begin{aligned}
\phi_1 &= \alpha(A)\alpha(B) & \phi_2 &= \alpha(A)\beta(B) \\
\phi_3 &= \beta(A)\alpha(B) & \phi_4 &= \beta(A)\beta(B)
\end{aligned} \tag{3-30}$$

If ψ is normalized then E in Eq. (3-28) is given by

$$E = \int \psi\mathcal{3C}\psi \, d\tau \tag{3-31}$$

The values of E are obtained by solving the determinant†

$$\begin{vmatrix}
\mathcal{3C}_{11} - E & \mathcal{3C}_{12} & \mathcal{3C}_{13} & \mathcal{3C}_{14} \\
\mathcal{3C}_{21} & \mathcal{3C}_{22} - E & \mathcal{3C}_{23} & \mathcal{3C}_{24} \\
\mathcal{3C}_{31} & \mathcal{3C}_{32} & \mathcal{3C}_{33} - E & \mathcal{3C}_{34} \\
\mathcal{3C}_{41} & \mathcal{3C}_{42} & \mathcal{3C}_{43} & \mathcal{3C}_{44} - E
\end{vmatrix} = 0 \tag{3-32}$$

† This is similar to the molecular orbital calculation procedure discussed in Chap. 2.

NUCLEAR MAGNETIC RESONANCE SPECTROSCOPY

Our next task is to evaluate the various $\mathcal{K}_{ij}$ terms in Eq. (3-32). To do this we employ the raising and lowering operators I^+ and I^- defined by

$$I^+ = I_x + iI_y$$
$$I^- = I_x - iI_y \qquad\qquad (3\text{-}33)$$

These operators have the following properties[3]

$$I^+\alpha = 0 \qquad I^+\beta = \alpha$$
$$I^-\alpha = \beta \qquad I^-\beta = 0 \qquad\qquad (3\text{-}34)$$

Thus we have

$$I^{A+}I^{B-} = (I_x{}^A I_x{}^B + I_y{}^A I_y{}^B) + i(I_y{}^A I_x{}^B - I_x{}^A I_y{}^B)$$
$$I^{A-}I^{B+} = (I_x{}^A I_x{}^B + I_y{}^A I_y{}^B) - i(I_y{}^A I_x{}^B - I_x{}^A I_y{}^B) \qquad (3\text{-}35)$$

so that

$$I_x{}^A I_x{}^B + I_y{}^A I_y{}^B = \tfrac{1}{2}(I^{A+}I^{B-} + I^{A-}I^{B+}) \qquad\qquad (3\text{-}36)$$

Hence Eq. (3-26) can be rewritten as

$$\mathcal{K} = -\nu_0(1 - \sigma_A)I_z{}^A - \nu_0(1 - \sigma_B)I_z{}^B$$
$$+ J[I_z{}^A I_z{}^B + \tfrac{1}{2}(I^{A+}I^{B-} + I^{A-}I^{B+})] \qquad (3\text{-}37)$$

It can be easily shown that if ϕ_i and ϕ_j have different values of M_I, we have

$$\int \phi_i \mathcal{K} \phi_j \, d\tau = \int \phi_j \mathcal{K} \phi_i \, d\tau = 0 \qquad\qquad (3\text{-}38)$$

The various $\mathcal{K}_{ij}$ terms in Eq. (3-32) are easily evaluated as

$$\mathcal{K}_{11} = -\frac{\nu_0}{2}(1 - \sigma_A) - \frac{\nu_0}{2}(1 - \sigma_B) + \frac{J}{4}$$

$$\mathcal{K}_{22} = \frac{\nu_0}{2}(1 - \sigma_A) - \frac{\nu_0}{2}(1 - \sigma_B) - \frac{J}{4}$$

$$\mathcal{K}_{33} = -\frac{\nu_0}{2}(1 - \sigma_A) + \frac{\nu_0}{2}(1 - \sigma_B) - \frac{J}{4}$$

$$\mathcal{K}_{44} = \frac{\nu_0}{2}(1 - \sigma_A) + \frac{\nu_0}{2}(1 - \sigma_B) + \frac{J}{4}$$

$$\mathcal{K}_{23} = \mathcal{K}_{32} = \frac{J}{2}$$

whereas $\mathcal{K}_{12}$, $\mathcal{K}_{21}$, $\mathcal{K}_{13}$, $\mathcal{K}_{31}$, $\mathcal{K}_{14}$, $\mathcal{K}_{41}$, $\mathcal{K}_{24}$, $\mathcal{K}_{42}$, $\mathcal{K}_{34}$, and $\mathcal{K}_{43}$ are all zero. Equation (3-32) now becomes

$$\begin{vmatrix} \dfrac{\nu_0}{2}(1-\sigma_A) + \dfrac{\nu_0}{2}(1-\sigma_B) + \dfrac{J}{4} - E & 0 & 0 & 0 \\[3mm] 0 & \dfrac{\nu_0}{2}(1-\sigma_A) - \dfrac{\nu_0}{2}(1-\sigma_B) - \dfrac{J}{4} - E & \dfrac{J}{2} & 0 \\[3mm] 0 & \dfrac{J}{2} & -\dfrac{\nu_0}{2}(1-\sigma_A) + \dfrac{\nu_0}{2}(1-\sigma_B) - \dfrac{J}{4} - E & 0 \\[3mm] 0 & 0 & 0 & -\dfrac{\nu_0}{2}(1-\sigma_A) - \dfrac{\nu_0}{2}(1-\sigma_B) + \dfrac{J}{4} - E \end{vmatrix} = 0$$

which immediately gives

$$E_1 = \nu_0\left(-1 + \frac{\sigma_A}{2} + \frac{\sigma_B}{2}\right) + \frac{J}{4} \tag{3-39a}$$

$$E_4 = \nu_0\left(1 - \frac{\sigma_A}{2} - \frac{\sigma_B}{2}\right) + \frac{J}{4} \tag{3-39b}$$

or simply

$$E_1 = -M + \frac{J}{4} \qquad E_4 = M + \frac{J}{4}$$

where

$$M = \nu_0\left(1 - \frac{\sigma_A}{2} - \frac{\sigma_B}{2}\right)$$

The other two solutions require that we find the roots for the determinant

$$
\begin{vmatrix}
\frac{\nu_0}{2}(1 - \sigma_A) & & \\
\quad - \frac{\nu_0}{2}(1 - \sigma_B) - \frac{J}{4} - E & & \frac{J}{2} \\
& & \\
\frac{J}{2} & & -\frac{\nu_0}{2}(1 - \sigma_A) \\
& & \quad + \frac{\nu_0}{2}(1 - \sigma_B) - \frac{J}{4} - E
\end{vmatrix} = 0
$$

We obtain

$$E = -\frac{J}{4} \pm \sqrt{\frac{\nu_0^2\delta^2}{4} + \frac{J^2}{4}}$$

where

$$\delta = \sigma_A - \sigma_B$$

We have†

$$E_2 = -\frac{J}{4} - \frac{1}{2}\sqrt{\nu_0^2\delta^2 + J^2}$$

$$\qquad = -\frac{J}{4} - \frac{1}{2}\nu_0\delta\left(1 + \frac{J^2}{2\nu_0^2\delta^2} + \cdots\right) \tag{3-39c}$$

and

$$E_3 = -\frac{J}{4} + \frac{1}{2}\sqrt{\nu_0^2\delta^2 + J^2}$$

$$\qquad = -\frac{J}{4} + \frac{1}{2}\nu_0\delta\left(1 + \frac{J^2}{2\nu_0^2\delta^2} + \cdots\right) \tag{3-39d}$$

† We use the series expansion $(1 + x)^{\frac{1}{2}} \simeq 1 + x/2 + x^2/4 + \cdots$ in Eqs. (3-39c) and (3-39d).

Figure 3-10 shows the change of energy levels due to the various interactions.

Having obtained the energy levels our next task is to obtain the selection rules. In order to induce transitions between levels m and n we need to apply an r-f field H_1 to the sample. Experimentally it is arranged so that H_1 oscillates in the x direction, which is perpendicular to the external magnetic field H_0 in the z direction. Applying H_1 along H_0 will not cause any transitions; the effect will merely be a modulation of the Zeeman energy levels. The interaction between H_1 and the nuclear magnetic moment is represented by $\mathfrak{JC}'$, which is given by

$$\mathfrak{JC}' = \mathbf{M}_x \cdot \mathbf{H}_1 \tag{3-40}$$

where $\mathbf{M}_x$ is the component of the nuclear dipole moment in the x direction. The transition probability is proportional to $[\int \psi_n (\mathbf{M}_x \cdot \mathbf{H}_1) \psi_m \, d\tau]^2$ or simply $(\int \psi_n I_x \psi_m \, d\tau)^2$ since $M_x = I_x \hbar \gamma$. It can be readily shown that by the substitution, $I_x = (I^+ + I^-)/2$, the last integral is zero unless the magnetic spin quantum numbers for the two states m and n differ by ± 1. Thus the selection rules are $\Delta M_I = \pm 1$.

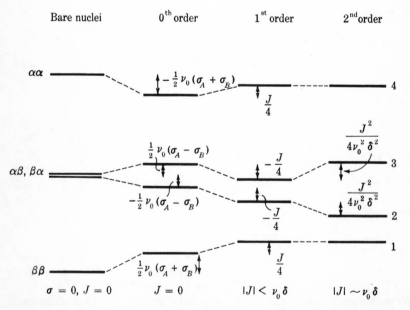

Fig. 3-10 The zeroth-, first-, and second-order perturbation of an AB system.

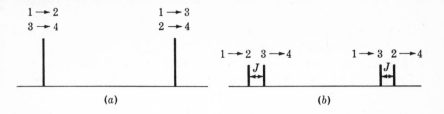

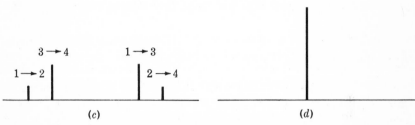

Fig. 3-11 Theoretical stick spectra of an AB system. (a) $J = 0$; (b) $|J| < \nu_0\delta$; (c) $|J| > \nu_0\delta$; (d) $\nu_0\delta = 0$.

The allowed transitions (in frequency) are:

$$E_2 - E_1 = -\frac{J}{2} + M - \frac{1}{2}\nu_0\delta\left(1 + \frac{J^2}{2\nu_0^2\delta^2} + \cdots\right)$$

$$E_3 - E_1 = -\frac{J}{2} + M + \frac{1}{2}\nu_0\delta\left(1 + \frac{J^2}{2\nu_0^2\delta^2} + \cdots\right)$$

$$E_4 - E_2 = \frac{J}{2} + M + \frac{1}{2}\nu_0\delta\left(1 + \frac{J^2}{2\nu_0^2\delta^2} + \cdots\right) \tag{3-41}$$

$$E_4 - E_3 = \frac{J}{2} + M - \frac{1}{2}\nu_0\delta\left(1 + \frac{J^2}{2\nu_0^2\delta^2} + \cdots\right)$$

Figure 3-11 shows the theoretical spectra for various values of J and $\nu_0\delta$.

Theoretical considerations show that J can be both positive and negative. However, the appearance of an NMR spectrum depends only on the absolute magnitude, and, hence, the modulus sign in Fig. 3-11.

It is important to note the following points in relation to Figs. 3-10 and 3-11.

1. In Fig. 3-10 the labeling of the levels ($\alpha\alpha$, $\alpha\beta$, etc.) applies only to the bare nuclei, zeroth- and first-order case. In the second-order perturbation case the levels originally corresponding to $\alpha\beta$ and $\beta\alpha$ are "mixed," and they should be represented by linear combinations of the type ($\alpha\beta \pm \beta\alpha$).

2. In the first-order case where $|J| < \nu_0\delta$, the system is said to be weakly coupled (ethanol at 40 MHz is an example) and is usually called an AX instead of the AB case. The AB case applies only when there is strong coupling between the nuclei, that is, $|J| > \nu_0\delta$.

3. Although the intensity and position of the lines may change depending on the magnitude of interaction, the total area under the experimental spectrum is the same for all cases.

4. When two nuclei have the same chemical shift, only one line is observed even though there is spin-spin interaction between them. This is why the NMR spectrum of benzene, for example, gives one line.

   In practice, the analysis of an NMR spectrum is largely done by trial and error. The procedure is to make intelligent guesses of J's and $\nu_0\delta$'s and construct theoretical spectra to compare with the experimental one. As Fig. 3-11 clearly shows, both J and $\nu_0\delta$ are most easily obtained from the first-order spectrum. Therefore, it is often advantageous to record spectra at the highest field possible so that most spectra will have the first-order appearance because of the increase in chemical shift. Table 3-2 gives typical J values for some organic compounds.

Table 3-2 Typical values of J for some organic compounds

Compound	J (Hz)
H—H	280
—CH_2—CH_3	7
$\begin{array}{c} \quad\ H \\ \diagdown\ \ \diagup \\ C{=}C \\ \diagup\ \ \diagdown \\ \quad\ H \end{array}$	1–5
$\begin{array}{c} H\quad\quad H \\ \diagdown\ \ \diagup \\ C{=}C \\ \diagup\ \ \diagdown \\ \quad\quad H \end{array}$	6–15
$\begin{array}{c} \quad\quad H \\ \diagdown\ \ \diagup \\ C{=}C \\ \diagup\ \ \diagdown \\ H \end{array}$	10–20
C_6H_6 (ortho)	5–9
(meta)	1–3
(para)	0–1

MAGNETIC EQUIVALENCE

Nuclei are said to be magnetically equivalent when they have the same chemical shift and the same spin-spin coupling constant with every other nucleus that does not have the same chemical shift.   Consider the following examples:

all protons equivalent

two protons not equivalent: $J_{HF}{}^{cis} \neq J_{HF}{}^{trans}$

none are equivalent: $J_{23} \neq J_{36}$, $J_{25} \neq J_{56}$, etc.

$$CH_3\!-\!\overset{\displaystyle H}{\underset{\displaystyle H}{\overset{|}{\underset{|}{C}}}}\!-\!OH$$

all methyl protons are equivalent if the rotation is rapid about the C—C bond

$$Br\!-\!\overset{\displaystyle F}{\underset{\displaystyle Cl}{\overset{|}{\underset{|}{C}}}}\!-\!\overset{\displaystyle H}{\underset{\displaystyle H}{\overset{|}{\underset{|}{C}}}}\!-\!I$$

the two protons are not equivalent regardless of the rate of rotation

LINESHAPE AND LINEWIDTH

There are two equations commonly employed to describe NMR lineshapes.[4]

Gaussian lineshape: $I(H) = \dfrac{T_2'}{\pi} \exp\left[-(H - H_m)^2\, \dfrac{T_2'^2}{\pi} \right]$ (3-42)

Lorentzian lineshape: $I(H) = \dfrac{T_2'}{1 + (H - H_m)^2 T_2'^2}$ (3-43)

where $I(H)$ is the intensity of absorption, H is the magnetic field in gauss or frequency, H_m is the field value at the maximum absorption, and T_2', which has the dimensions of time, is called the linewidth parameter.   These lineshape plots are shown in Fig. 3-12.   An approximate relation for T_2' is

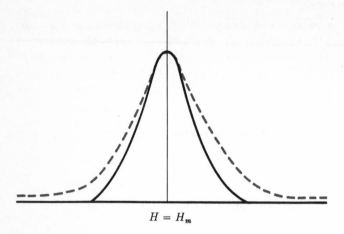

$H = H_m$

Fig. 3-12 The gaussian (solid line) and Lorentzian (dotted line) lineshape of an absorption line.

given by

$$\frac{1}{T'_2} = \frac{1}{T_1} + \frac{1}{T_2} \tag{3-44}$$

where T_1 is the *nuclear-spin-relaxation* time and T_2 the *spin-spin* or *transverse relaxation* time.   A discussion of T_1 and T_2 is given in Appendix 5.

NUCLEI HAVING SPIN $I > \frac{1}{2}$

Nuclei such as H^2, N^{14}, Cl^{35}, etc., having $I > \frac{1}{2}$ do not have spherical symmetry of charge distribution and possess electric quadrupole moment eQ, where Q is a measure of the deviation of the charge distribution from spherical symmetry.   Figure 3-13 shows the two possible orientations of the charge distributions.   Spin multiplets are often broadened or even collapsed by nuclei causing the splittings if these nuclei possess an electric

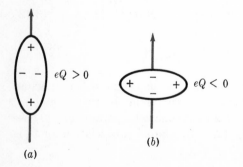

Fig. 3-13 Nuclear charge distribution for nuclei having $I > \frac{1}{2}$.   (a) Prolate; (b) oblate.   The positive and negative signs indicate only the relative charge distributions.

quadrupole moment. For example, the proton magnetic resonance spectrum of $N^{15}H_3$ (N^{15}, $I = \frac{1}{2}$) consists of two sharp lines of equal intensity as a result of the spin-spin interaction with the N^{15} nucleus. The same spectrum of $N^{14}H_3$ (N^{14}, $I = 1$) shows the expected but also very much broadened three lines. The reason for this line-broadening phenomenon is as follows: The electric quadrupole moment of the nucleus in question interacts with the fluctuating electric field gradients which are present in asymmetric chemical bonds (see Chap. 4), and this causes a reorientation of the quadrupole in the electric field. Since the quadrupole moment possesses a fixed orientation with respect to the nuclear spin axis and therefore the nuclear magnetic moment, its reorientation will be accompanied by a reorientation of the magnetic moment with respect to the applied magnetic field. This effectively shortens the spin-lattice relaxation time T_1 and consequently the spin multiplets caused by the nuclei will be broadened or collapsed. The proton magnetic resonance of NH_4^+, however, contains much sharper lines. Because of the tetrahedral symmetry arrangement of four identical protons, there is no asymmetry in the bonds.[5]

3-4 APPLICATIONS

Structural determination Although the NMR technique is now being employed to solve various chemical and biochemical problems, structural determination still ranks foremost among these. The elucidation of molecular structure from NMR spectra is essentially the determination of two quantities, that is, chemical shift and spin-spin coupling constant. The former tells us about the nature of the functional groups and the latter about the arrangement of the atoms. Furthermore, the relative intensity of the peaks tells us about the ratio of the same atoms in different chemical environments. However, the task of analyzing spectra of large molecules can be quite complex because of the spin-spin interaction. It is indeed fortunate that naturally abundant carbon (C^{12}) and oxygen (O^{16}) do not have nuclear magnetic moments, otherwise structural studies would be limited to only very simple molecules.

Complex spectra analysis can often be aided by either chemical or physical means. We could, for example, replace some hydrogen atoms with deuteriums and compare their spectra. Since isotopic substitution has a negligible effect on the electronic structure of the molecule, the chemical shifts will remain unchanged but the coupling constants will be different because deuterons have a smaller magnetic moment. Since deuteron has a spin $I = 1$, we might expect that the resulting spectrum of the deuterated compound would be more complex. However, spin-spin interactions due to deuterons are seldom observed because of the quad-rupole relaxation effect discussed earlier.

Although deuterium substitution can be of great aid, it is not always easy to synthesize deuterated compounds.   Also, in cases where deuteron quadrupole relaxations are not sufficient to completely collapse the spin multiplets, broadened lines will be obtained.   An experimental technique which can greatly simplify spectra without these complications is the *double resonance* method.   This involves the simultaneous irradiation of the sample with two different r-f fields.   The frequency of one field is adjusted for nuclei whose spectra are being recorded, and the other is adjusted to cause *saturation* of the nuclei which are causing the splittings. The saturation of the nuclei causes the effect of spin-spin interaction to disappear, and the system is equivalent to the hypothetical compound in which certain nuclei are replaced with isotopes having a nuclear spin quantum number $I = 0$.   This technique is also known as spin decou-pling.   The applicability of this method depends on the stability of the instrument.   We can have either homogeneous or heterogeneous spin decoupling.   As an example of the latter let us consider the single and

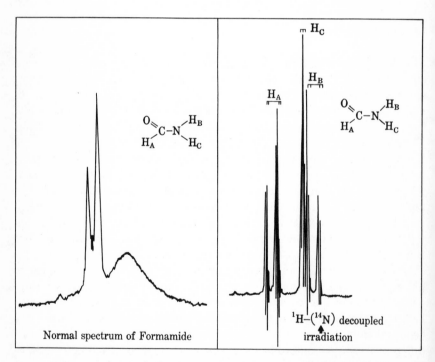

Fig. 3-14 The normal and N^{14} decoupled NMR spectra of *N,N*-formamide. (*By permission from JEOLCO, Inc.*)

double resonance spectra of N,N-formamide (Fig. 3-14). This compound contains three inequivalent protons strongly coupled together and is known as an ABC case. The broad peaks in Fig. 3-14a are due to the nitrogen quadrupole relaxation. The coupling constants shown in Fig. 3-14b are the proton spin-spin interactions according to the structure

The values obtained are $J_{AB} = 13$ Hz, $J_{AC} = 2.1$ Hz, and J_{BC} is very small, probably of the order of the linewidth.[6]

As Eq. (3-22) shows, it is possible to evaluate internuclear distances from the splittings of the NMR lines in solids. This technique is not generally applicable, however, because dipolar interactions due to the nuclei in neighboring sites often broaden lines and obscure the splittings. A technique developed since 1963 of using liquid crystals as solvent has greatly increased the use of NMR in the determination of molecular parameters. Liquid crystals are compounds which when melted possess certain properties over a limited range of temperatures found only in crystalline solids. Thus when p,p'-di-n-hexyloxyazoxybenzene, a typical liquid crystal, melts at 118°C, it forms a liquid in which all the molecules are aligned with their long axes parallel. This state persists through 135°C, where there is a second phase transition and the liquid becomes isotropic. NMR studies have been carried out by dissolving the compound in a suitable liquid crystal at the temperature at which the solvent shows the liquid crystal properties. The alignment of the solute molecules will usually be similar to the alignment of the solvent molecules. Since the solute molecules can diffuse as rapidly as they do in the normal isotropic liquid phase, most of the intermolecular dipolar interactions will vanish. On the other hand, because of the orientation along the long axes, the intramolecular dipolar interactions remain. Consequently the NMR spectrum will consist of the fine structures due to both spin-spin and dipolar interactions but the linewidths are comparable to those normally obtained in isotropic liquids. In this manner accurate bond lengths and, indirectly, bond angles have been determined for a number of compounds. For example, the NMR study of cyclopropane in liquid crystal has yielded the following properties: $C—C = 1.123$ Å and $\widehat{HCH} = 114.4°$.[7]

Hindered rotation about a single bond   The ABC case of formamide discussed above arises because the protons attached to the nitrogen are not equivalent. Rotation about the C—N bond is hindered because of the partial double-bond character.   In general we write

$$
\begin{array}{ccc}
R \qquad R & & R \qquad R \\
\diagdown \quad \diagup & & \diagdown \quad \diagup \\
C{-}N & \leftrightarrow & C{=}N^{+} \\
\diagup\diagup \qquad \diagdown & & \diagup \qquad \diagdown \\
O \qquad\quad R & -O & \qquad R
\end{array}
$$

where R = alkyl, H.   If the rate of rotation about the C—N bond measured in frequency is less than the difference in the chemical shifts of the
protons also measured in frequency, two magnetically different groups of
protons will result.   Magnetic equivalence of these protons may be
achieved by increasing the rate of rotation, that is, by heating the compound.   By following the changes of the NMR spectrum over a range of
temperatures, the energy of activation ΔE for the rotation can be
calculated.   A well-studied example is N,N'-dimethylacetamide (DMA).[8]

$$
\begin{array}{ccc}
CH_3 & & CH_3 \\
\diagdown & & \diagup \\
& C{-}N & \\
\diagup\diagup & & \diagdown \\
O & & CH_3
\end{array}
$$

There is no spin-spin coupling between protons on different methyl groups
because of the large separations.   At $-30°C$ the spectrum shows a
single peak and a doublet indicating the inequivalence of the methyl
groups attached to the nitrogen.   As the temperature is increased the
peaks of the doublet begin to broaden and eventually collapse into a
single line at about 90°C.   The rate of rotation k is given by

$$ k = \frac{1}{2\tau} $$

where τ is the mean lifetime of a particular configuration. τ is related to
the linewidth and line separation of the methyl peaks and can be measured
at any temperature using the appropriate equations.   The activation
energy ΔE can then be calculated using the Arrhenius equation

$$ \ln k = \ln A - \frac{\Delta E}{RT} $$

where A is a constant characteristic of the reaction.   For DMA, ΔE is
about 12 kcal mole^{-1}.

Intermolecular exchange reactions   In discussing the high-resolution
spectrum of ethanol earlier there was no mention of the spin-spin interaction between the hydroxyl proton and the $-CH_2$ group.   Such a
spectrum arises whenever there is a trace of acid or base present because

of the following reactions:

$$C_2H_5OH + H_3O^+ \overset{k_1}{\rightleftharpoons} C_2H_5OH_2^+ + H_2O \tag{1}$$

$$C_2H_5OH' + C_2H_5OH_2^+ \overset{k_2}{\rightleftharpoons} C_2H_5OHH'^+ + C_2H_5OH \tag{2}$$

$$C_2H_5OH + OH^- \overset{k_3}{\rightleftharpoons} C_2H_5O^- + H_2O \tag{3}$$

$$C_2H_5OH + C_2H_5O^- \overset{k_4}{\rightleftharpoons} C_2H_5O^- + C_2H_5OH \tag{4}$$

If a nucleus undergoes exchange between two chemically and hence magnetically different sites, the resonance of the nucleus in these two sites will be broadened if the frequency of exchange is of comparable magnitude to the difference in chemical shifts. This follows from the Heisenberg uncertainty principle

$$\Delta E \, \Delta t \sim \hbar$$
$$\Delta E = h \, \Delta\nu$$

hence

$$\Delta\nu = \frac{1}{\Delta t}$$

The uncertainty in frequency (width of the resonance line at half-height) is approximately equal to the reciprocal of the uncertainty in time for which the nucleus has a given energy, which is the mean lifetime τ. That is

$$\Delta t = \tau$$

The broadening of the hydroxyl lines in ethanol proceeds roughly as follows

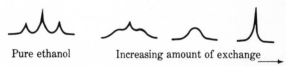

Pure ethanol   Increasing amount of exchange

As the rate of exchange increases, the —OH triplet due to the spin-spin splitting by the —CH$_2$ group begins to broaden and finally coalesces into a single sharp line. The —CH$_2$ lines are similarly affected by the exchange process. At very high exchange rates all the spin multiplets of the hydroxyl protons disappear because the lifetime of the exchanging proton in a given site is too short to establish spin-spin interaction. The —CH$_3$ group is not affected by this process. The evaluation of rate constants from line-broadening measurements is fairly straightforward for molecules having simple NMR spectra. For the ethanol case we get $k_1 = 2.8 \times 10^6$ l mole^{-1} sec^{-1}, $k_2 = 1.1 \times 10^6$ l mole^{-1} sec^{-1}, $k_3 = 2.8 \times 10^6$ l mole^{-1} sec^{-1}, and $k_4 = 1.4 \times 10^6$ l mole^{-1} sec^{-1}.[9]

A rather interesting feature of this kind of kinetic study is that all measurements are carried out without disturbing the chemical equilibria whatsoever.

PROBLEMS

3-1. Calculate the Maxwell-Boltzmann distribution for protons having $I = \frac{1}{2}$ at 300°K and in a field of 13,500 G.

3-2. If an NMR experiment is performed using a field of 13,500 G, at what frequency would the resonance occur for the following nuclei: H^2, H^2, C^{13}, N^{14}, O^{16}, O^{17}, F^{19}, and P^{31}?

3-3. Sketch the NMR spectrum of isopropyl alcohol given the chemical shifts of the —CH_3, —CH, and —OH protons as 70, 240, and 95 Hz, respectively (downfield from TMS), and the spin coupling constant between the —CH_3 and —CH protons as 6 Hz. Indicate the relative intensity of the lines in the first-order approximation.

3-4. The protons in 2-bromo-5-chlorothiophene are a good example of an AB spin system. At 30.5 MHz the NMR spectrum is that shown below:

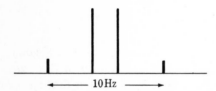

The chemical shift and coupling constant for the 30.5-MHz spectrum are $|\nu_0\delta| = 4.7$ Hz and $|J| = 3.9$ Hz.

(*a*) Use the equations derived in this chapter for the AB system to calculate the positions of the lines in the 30.5-MHz spectrum with respect to the center.

(*b*) Calculate the positions that would be expected if the spectrum were recorded at 60 MHz.

3-5. Calculate the relative intensity of transition lines for the AB case.

SUGGESTIONS FOR FURTHER READING

Introductory

Hecht, H. G.: "Magnetic Resonance Spectroscopy," John Wiley & Sons, Inc., New York, 1967.

Jackman, L. M.: "Applications of Nuclear Magnetic Resonance Spectroscopy," Pergamon Press, New York, 1959.

Roberts, J. D.: "Nuclear Magnetic Resonance," McGraw-Hill Book Company, New York, 1959.

Intermediate-advanced

Carrington, A., and A. D. McLachlan: "Introduction to Magnetic Resonance," Harper & Row, Publishers, Incorporated, New York, 1967.

Emsley, J. W., J. Feeney, and L. H. Sutcliffe: "High Resolution Nuclear Magnetic Resonance Spectroscopy," Pergamon Press, Ltd., London, 1966.

Jackman, L. M., and Sternhell, S.: "Applications of Nuclear Magnetic Resonance Spectroscopy in Organic Chemistry," Pergamon Press, New York, 1969.
Pople, J. A., W. G. Schneider, and H. J. Berstein: "High-resolution Nuclear Magnetic Resonance," McGraw-Hill Book Company, New York, 1959.

READING ASSIGNMENTS

Measurements Based on the Symmetry of NMR Spectra, J. M. Anderson, *J. Chem. Educ.*, **42**: 363 (1965).
NMR Spectra: Appearance of Patterns from Small Spin Systems, E. D. Beaker, *J. Chem. Educ.*, **42**: 591 (1965).
Nuclear Magnetic Resonance, F. A. Borey, *Chem. Eng. News*, **43**: 98 (1965).
Analysis of Complex NMR Spectra for the Organic Chemists, E. W. Garbisch, Jr., *J. Chem. Educ.*, **45**: 311, 402, and 480 (1968).
Magnetic Resonance Spectra of Multispin Systems, P. L. Corio and R. C. Hirst, *J. Chem. Educ.*, **46**: 345 (1969).
Calculation of NMR Shift Using Particle in a Box Wave Functions, R. B. Flenwelling and W. G. Laidlaw, *J. Chem. Educ.*, **46**: 355 (1969).
Magnetic Double-resonance Techniques in Chemistry, W. McFarlane, *Chem. Britain*, **5**: 142 (1969).
A Proton Magnetic Resonance Coordination Number Study, A. Fratiello and R. E. Schuster, *J. Chem. Educ.*, **45**: 91 (1968).
Kinetics of Proton Exchange of Trimethylammonium Ion by NMR, D. E. Leyden and W. R. Morgan, *J. Chem. Educ.*, **46**: 169 (1969).
An NMR Determination of Optical Activity, J. Jacobus and M. Raban, *J. Chem. Educ.*, **46**: 351 (1969).
An NMR Laboratory-Problem for Introductory Quantum Chemistry, K. F. Kuhlmann and C. L. Braun, *J. Chem. Educ.*, **46**: 750 (1969).
Nuclear Magnetic Resonance Studies of Molecular Motion in Polymers, W. P. Slichter, *J. Chem. Educ.*, **47**: 193 (1970).

REFERENCES

1. For derivation of Eq. (3-19), see Bersohn and Baird, "An Introduction to Electron Paramagnetic Resonance," p. 208, W. A. Benjamin, Inc., New York, 1966.
2. G. E. Pake, *J. Chem. Phys.*, **16** (1948).
3. For derivation of Eq. (3-34), see Carrington and McLachlan, p. 246.
4. For a general discussion of magnetic resonance lineshapes, see L. Petrakis, *J. Chem. Educ.*, **44**: 432 (1967).
5. R. A. Ogg, Jr., and J. D. Ray, *J. Chem. Phys.*, **26**: 1339, 1515 (1957).
6. L. H. Piette, J. D. Ray, and R. A. Ogg, Jr., *J. Mol. Spectry.*, **2**: 66 (1958).
7. L. C. Snyder and S. Meiboom, *J. Chem. Phys.*, **47**: 1480 (1967). For a general discussion of liquid crystals as solvents in NMR, see G. R. Luckhurst, *Quart. Rev.*, **22**: 179 (1968).
8. M. T. Rogers and J. C. Woodbrey, *J. Phys. Chem.*, **66**: 540 (1962).
9. Z. Luz, D. Gill, and S. Meiboom, *J. Chem. Phys.*, **30**: 1540 (1958). For a comprehensive discussion of chemical rate processes and magnetic resonance, see C. S. Johnson, Jr., in J. S. Waugh (ed.), "Advances in Magnetic Resonance," vol. 1, p. 33, Academic Press, Inc., New York, 1965.

4
Nuclear Quadrupole Resonance Spectroscopy

4-1 INTRODUCTION

As we saw in Chap. 3, nuclei having spin $I > \frac{1}{2}$ have a nonspherical distribution of charge density and therefore possess an electric quadrupole moment eQ.† Q is a measure of the deviation of the nuclear charge from spherical symmetry and is given by[1]

$$Q = \int \rho r^2 (3 \cos^2 \theta - 1) \, d\tau \tag{4-1}$$

where ρ is the nuclear charge density, r is the distance from the origin to the element $d\tau$, and θ is the angle between r and the spin axis. If the nucleus possessing a quadrupole moment is placed in an inhomogeneous electric field, the potential energy of the quadrupole will vary depending on the orientation of the quadrupole moment with the field (Fig. 4-1).

Analogous to the NMR case, quantum mechanics imposes the restriction that there can be only certain orientations of the spin axis in space.

† Electric quadrupole moment is sometimes simply expressed as Q.

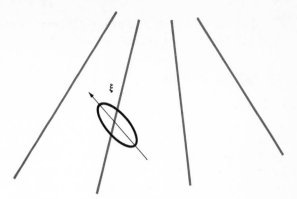

Fig. 4-1 A prolate nucleus in an inhomogeneous electric field.

In an inhomogeneous electric field, the interaction is between the nuclear quadrupole moment and the electric field gradient (EFG) q at the nucleus. In the cartesian coordinates we define the EFGs along the x, y, and z axes as

$$q_{xx} = \frac{\partial^2 V}{\partial x^2} \tag{4-2a}$$

$$q_{yy} = \frac{\partial^2 V}{\partial y^2} \tag{4-2b}$$

$$q_{zz} = \frac{\partial^2 V}{\partial z^2} \tag{4-2c}$$

where V is the potential at the nucleus due to some electric charge. In atoms and molecules the EFG at a nucleus arises as a result of the surrounding electrons. Depending on the particular symmetry and electronic configuration, we can have any one of the following three situations.

1. $q_{xx} = q_{yy} = q_{zz}$

This corresponds to the special symmetry of the field gradient; all orientations of the quadrupole moment with the field have the same energy; consequently the degeneracy corresponding to M_I is not removed, and no resonance can occur. An example of this is the chloride ion Cl^-. The closed-shell structure has a spherical symmetry and the EFG at the nucleus along the three axes are the same.

2. $q_{xx} = q_{yy} \neq q_{zz}$

This corresponds to the axial symmetry along the z axis. For example, in the FCl molecule, the EFG at the chlorine nucleus along the

F—Cl bond, defined as the z axis, is different from those along the x and y axes. According to the Laplace equation

$$\frac{\partial^2 V}{\partial x^2} + \frac{\partial^2 V}{\partial y^2} + \frac{\partial^2 V}{\partial z^2} = 0 \tag{4-3}$$

we have

$$q_{zz} = -2q_{xx} = -2q_{yy} \tag{4-4}$$

for axial symmetry. Since any two of the field gradients are determined by the third, it is only necessary to use one parameter to define the field inhomogeneity. The standard convention is to choose q_{zz}, and the energy of interaction between the quadrupole and the field gradient is given by

$$E = eQq_{zz}f(\xi) \tag{4-5}$$

where E is proportional to a function $f(\xi)$ of the angle ξ. Frequently q_{zz} is simply replaced by q so that

$$E = eQqf(\xi) \tag{4-6}$$

The energy levels for a nucleus having spin I corresponding to this are given by

$$E = eQq \left[\frac{3M_I^2 - I(I + 1)}{4I(2I - 1)} \right] \tag{4-7}$$

where eQq is the nuclear quadrupole coupling constant. Figure 4-2 shows

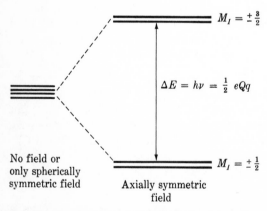

Fig. 4-2 Energy levels arising from the interaction of a nucleus with $I = \frac{3}{2}$ with an axially symmetric field.

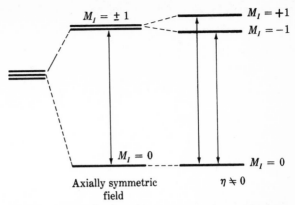

Fig. 4-3 Energy levels arising from the interaction of a nucleus with $I = 1$ with an axially symmetric and an axially nonsymmetric field.

the splitting of the nuclear quadrupole energy levels for a nucleus having $I = \frac{3}{2}$. Because of the M_I^2 term in Eq. (4-7), only one line is observed. The resonance condition is

$$\Delta E = h\nu = \tfrac{1}{2}eQq \tag{4-8}$$

For most nuclei, ν is in the megahertz range, and for this reason nuclear quadrupole resonance (NQR), like NMR, is a branch of r-f spectroscopy.

3. $q_{xx} \neq q_{yy} \neq q_{zz}$

This is the most complex case. The energy of interaction is given by

$$E = \frac{eQq}{4I(2I - 1)} [3M_I^2 - I(I + 1)] \left(1 + \frac{\eta^2}{3}\right)^{\frac{1}{2}} \tag{4-9}$$

where η, the asymmetry parameter, is given by

$$\eta = \frac{q_{xx} - q_{yy}}{q_{zz}} \tag{4-10}$$

The M_I degeneracy of the energy levels is removed for nuclei having integral spins if $\eta \neq 0$. For half-integral spins the M_I degeneracy is unaffected whether η is zero or not. Figure 4-3 shows the splitting of the quadrupole energy levels for $I = 1$. An example of $\eta \neq 0$ is the chlorine nucleus in $CH_2{=}CHCl$.

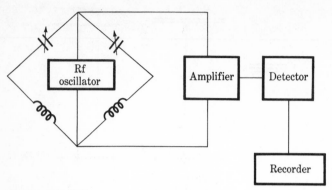

Fig. 4-4 Block diagram of an NQR spectrometer. The sample tube (not shown) is placed in one of the coils.

4-2 EXPERIMENTAL TECHNIQUES

Unlike the NMR experiment where an *external* magnetic field is required to remove the M_I degeneracy, the electric field required in an NQR experiment is provided by the electrons in the molecule.   This means that the transition frequencies are fixed by the electronic structure of the molecule, and we observe the resonance by varying the frequency instead of the field.   Figure 4-4 shows the schematic diagram of an NQR spectrometer. Power from a r-f oscillator is supplied to a previously balanced bridge circuit.   At resonance, energy is absorbed by the sample in the inductive coil and this causes an imbalance in the circuit.   The resultant output voltage is then amplified and recorded.   Spectra obtained this way are often called the pure quadrupole resonance spectra since the electric quadrupole constant eQq can also be obtained from the fine structure of the rotational spectra (see Chap. 7).

NQR spectra can be obtained only for solids.   In liquids and gases, at normal pressure, collision between molecules changes the axis of rotation continuously, and consequently the EFG at the nucleus will be averaged to zero.   Because of this limitation, NQR does not have the same range of applications as NMR.

4-3 THEORY

NUCLEAR QUADRUPOLE COUPLING IN ATOMS

In an atom the EFG at the nucleus is due to the electrons in the various valence shells.   Closed shells have spherical symmetry and do not contribute to the field gradient.† The same is true for electrons in the s orbitals. Thus only the p and d electrons in the valence shells contribute to the EFG. For an atom having one valence electron, the EFG, q_{atom}, is given by

$$q_{atom} = e \int \phi^* \left(\frac{3 \cos^2 \theta - 1}{r^3} \right) \phi \, d\tau \tag{4-11}$$

where ϕ is the electronic wave function and θ the angle between the fixed z axis in space and the radius vector r (from the nucleus to the electron). ϕ can be separated into the angular and radial parts with the result that

$$q_{atom} = - \frac{2le}{2l + 3} \left(\frac{1}{r^3} \right)_{av} \tag{4-12}$$

where l is the orbital angular momentum of the electron and $(1/r^3)_{av}$ is the average distance between the nucleus and the electron.

The quantity of prime interest is, of course, the EFG, since it tells us about the electronic structure of the atom. Nuclear quadrupole coupling constants of a number of atoms have been measured by the atomic-beam technique, and the EFGs for atoms of known quadrupole moment Q have been evaluated. They agree quite well with the calculated values.

NUCLEAR QUADRUPOLE COUPLING IN MOLECULES

In a molecule the EFG at a particular nucleus due to the valence electrons is given by

$$q_{mol} = \int \psi^* \left(\frac{3 \cos^2 \theta - 1}{r^3} \right) \psi \, d\tau \tag{4-13}$$

where ψ is the molecular wave function of the electrons. According to the theory of Townes and Dailey,[2] the molecular wave function can be expanded in terms of the atomic wave functions

$$\psi = \sum_i a_i \phi_i \tag{4-14}$$

They showed that the dominant contributions of EFG at a nucleus are due to the p-type valence electrons associated with the chemical bonding as well as p-type lone-pair electrons. Contributions due to other charges in the molecule and the various polarization effects on the inner shells are all negligible. In a molecule, the electronic structure near a nucleus depends on the hybridization of the bonding orbital and the ionic character of the bond. The equation relating the nuclear quadrupole constants

† This is not *strictly* true because of the polarization effects. However, these effects are small and can often be neglected.

Table 4-1   Nuclear quadrupole coupling constants of
Cl^{35}, Br^{79}, and I^{127}, and some of their compounds†

Molecule	eQq_{mol} (MHz)	Molecule	eQq_{mol} (MHz)
Cl (atomic)	-109.7	Br (atomic)	769.7
BrCl	-103.6	BrCl	876.8
ICl	-82.5	LiBr	37.2
FCl	-145.9	NaBr	58
KCl	0.04	KBr	10.2
RbCl	0.77	CH_3 Br	528.9
CsCl	3	I (atomic)	-2292.8
CH_3Cl	-74.7	NaI	-259.9
		KI	-60

† Values taken, with permission, from C. H. Townes and
A. L. Schawlow, "Microwave Spectroscopy," McGraw-
Hill Book Company, New York, 1955.

eQq_{mol} of a halogen atom to that measured in an isolated atom eQq_{atom} is
given by

$$eQq_{mol} = [1 - s + d - i(1 - s - d)]eQq_{atom} \qquad (4\text{-}15)$$

where s and d denote the amount of s and d character of the bonding orbital
and i is the ionic character of the bond.†   When there is also π bond
present, Eq. (4-15) is modified to give

$$eQq_{mol} = (1 - s + d - i - \pi)eQq_{atom} \qquad (4\text{-}16)$$

Table 4-1 shows the eQq_{mol} and eQq_{atom} values for some halogen compounds.

A comparison of eQq_{mol} and eQq_{atom} tells us about the nature of the
bond.   For example, eQq_{mol} of chlorine in KCl is almost zero, indicating
that the molecule must be essentially ionic.

4-4   APPLICATIONS

Nature of chemical bond   NQR is most frequently employed to investigate the electronic structure of molecules.   As Eqs. (4-15) and (4-16)
show, a comparison of the nuclear quadrupole coupling constant in the
atomic and molecular state for the same nucleus provides information
regarding the extent of hybridization and the ionic character of the bond.

† If ψ is the bonding orbital for the AB molecule, then

$$\psi = a\phi_A + b\phi_B$$

and the ionic character i is defined as $i = a^2 - b^2$, if A is more electronegative than B.

Unfortunately there are usually too many unknowns involved, and it is not often possible to deduce the exact nature of bonding simply from the NQR data alone.   Thus the same data are often interpreted differently by different scientists.   Here we shall discuss a more clear-cut example— the bonding in H_2S.   The electronic configuration of sulfur is $[Ne]3s^23p^4$ and because $\widehat{HSH}$ in H_2S is nearly 90° it had been assumed for many years that pure p orbitals of sulfur were involved in the bonding.   However, the NQR measurement on H_2S showed a large asymmetry parameter of $\eta = 0.60$ for sulfur.[3]   In an asymmetric molecule the EFG along the three axes can be written as

$$eQq_{xx} = -\left(\frac{N_y + N_z}{2} - N_x\right)eQq_{\text{atom}}$$

$$eQq_{yy} = -\left(\frac{N_x + N_z}{2} - N_y\right)eQq_{\text{atom}}$$

$$eQq_{zz} = -\left(\frac{N_x + N_y}{2} - N_z\right)eQq_{\text{atom}}$$

where N_x, N_y, and N_z are the effective electron populations of the p_x, p_y, and p_z orbitals of the atom.   Hence η can be written as

$$\eta = \frac{q_{xx} - q_{yy}}{q_{zz}} = \frac{3(N_y - N_x)}{N_x + N_y - 2N_z} = -0.60$$

If we let the p_z orbital contain the lone pair, we have $N_z = 2$ and $1.33N_y = N_x + 1$.   It is clear that N_x cannot be equal to N_y, and therefore we do not have pure p-type bonding.   A possible explanation for the large asymmetry parameter observed is that the bonding orbitals have about 15 percent d character.

Structural information   A number of group III halides of the type MX_3 (for example, $AlBr_3$) have been studied by the NQR technique. The basic halogen spectrum consists of three resonance lines: two are closely spaced together and are far above the third.   The fact that different resonance frequencies are observed for the same nuclei indicates that they are not chemically equivalent.   This is supported by the x-ray data that these halides exist as dimers and there are two types of halogen atoms corresponding to the bridge and end positions

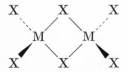

The small difference of one set of resonance frequencies is thought to be due to a slight difference of the crystalline field around the chemically equivalent set of atoms.

Study of charge-transfer compounds   Charge-transfer compounds (see Chap. 11) are believed to be formed when equimolar CBr_4 and p-xylene are brought together.   The same is true for mixtures of CCl_4 and p-xylene and Br_2 and benzene.   Hooper has studied the frozen solutions of these mixtures and found that the resonance frequencies of the halogens are not appreciably different from those in the pure molecular state.[4]   The conclusion therefore is that there is a lack of "charge transfer" of these compounds in the ground state.

PROBLEMS

4-1. Draw energy level diagrams and calculate transition frequencies for a nucleus having $I = \frac{3}{2}$ and assuming (a) $\eta = 0$ and (b) $\eta \neq 0$.

4-2. The bonding in a series of linear trihalide anions of the type IX_2^- (where $X = Cl$, Br, and I) is not well established and there are two theoretical interpretations: (a) The central iodine atom is sp^3d-hybridized (trigonal bipyramid); the two equatorial hybrid orbitals are used to form bonds with X while the three axial ones are used to accommodate the lone pairs.   (b) The chemical bonds are formed between iodine and X simply by the overlap of the p orbitals.

How would you employ NQR to investigate the nature of bonding in these compounds?

4-3. Gordy has suggested the use of $1 - eQq_{mol}/2eQq_{atom}$ as a measure of the ionic character of diatomic molecules containing halogens.   Plot this quantity vs. the electronegativity difference between the two atoms.

SUGGESTIONS FOR FURTHER READING

Introductory

W. J. Orville-Thomas, *Quart. Rev.*, **11**: 162 (1957).

Intermediate-advanced

Das, T. P., and E. L. Hahn: "Nuclear Quadrupole Resonance Spectroscopy," Academic Press, Inc., New York, 1958.

Gordy, W.: Chemical Applications of Spectroscopy, in W. West and A. Weissberger (eds.), "Techniques of Organic Chemistry," vol. 9, Interscience Publishers, Inc., New York, 1956.

Lucken, E. A. C.: "Nuclear Quadrupole Coupling Constants," Academic Press, Inc., New York, 1969.

Townes, C. H., and A. L. Schawlow: "Microwave Spectroscopy," McGraw-Hill Book Company, New York, 1955.

READING ASSIGNMENTS

Nuclear Quadrupole Resonance, R. S. Drago, *Anal. Chem.*, **38**: 31A (1966).
Analytical Aspects of Nuclear Quadrupole Resonance Spectrometry, E. G. Brame, Jr.,
  Anal. Chem., **39**: 918 (1967).

REFERENCES

1. For derivation of Eq. (4-1), see Townes and Schawlow, p. 133.
2. C. H. Townes and B. P. Dailey, *J. Chem. Phys.*, **17**: 782 (1949).
3. C. A. Burrus and W. Gordy, *Phys. Rev.*, **92**: 274 (1963).
4. H. O. Hooper, *J. Chem. Phys.*, **41**: 599 (1964).

5
Mössbauer Spectroscopy

5-1 INTRODUCTION

Both NMR and NQR are concerned with the ground-state properties of the nucleus. However, there is another kind of spectroscopy that deals with both the excited- as well as the ground-state properties. Consider the following decay scheme of Co^{57} to Fe^{57} shown in Fig. 5-1. The excited Co^{57} nucleus decays to the stable Fe^{57} nucleus through electron capture, followed by emission of delayed γ-rays, and this latter phenomenon is known as γ-ray fluorescence. In the presence of a target nucleus, that is, Fe^{57}, the emitted radiation can be resonantly absorbed. Because of the finite lifetime τ associated with the excited state, there is an uncertainty in the energy of the emitted γ-rays. If we define Γ to be the linewidth of the emitted line, we have

$$\Gamma \tau \sim \hbar \tag{5-1}$$

Nuclear fluorescence experiments usually involve the low-lying excited states having lifetimes ranging from 10^{-11} to 10^{-4} sec. According

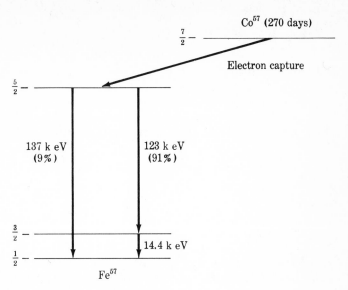

Fig. 5-1 Decay scheme of Fe⁵⁷.

to Eq. (5-1), therefore, this would give 10^{-4} to 10^{-11} eV as the corresponding range for the linewidths. Alternatively, Eq. (5-1) can be written as

$$\Gamma = \frac{\hbar \ln 2}{\tau_{\frac{1}{2}}} \tag{5-2}$$

where $\tau_{\frac{1}{2}}$ is the half-life of the excited state.

In a nuclear fluorescence experiment, the emitted γ-ray quanta
carry away momentum p given by

$$p = \frac{h}{\lambda} \tag{5-3}$$

where λ is the wavelength of the radiation. Conservation of linear
momentum requires that the nucleus recoil with an equal momentum in
the opposite direction. The corresponding recoil kinetic energy R of
the nucleus is given by

$$R = \frac{p^2}{2M} \tag{5-4}$$

where M is the mass of the recoiling nucleus. Similarly the target
nucleus upon receiving the γ-ray must also recoil with energy R. Consequently some of the energy of the γ transition is converted into recoil
energy. If E_0 is the energy above the ground state, we have, for the

emitting nucleus,

$$E = E_0 - \frac{p^2}{2M} \tag{5-5}$$

and for the absorber

$$E = E_0 + \frac{p^2}{2M} \tag{5-6}$$

Thus the emission and the absorption lines will be centered $2R$ apart. Since

$$E_0 = h\nu = h\frac{c}{\lambda} = pc \tag{5-7}$$

we have†

$$2R = \frac{E_0{}^2}{Mc^2} \tag{5-8}$$

In order to have resonance absorption, the linewidth Γ must be greater than or equal to the loss in γ-ray energy due to recoil, that is, we must have

$$2R \sim \Gamma \tag{5-9}$$

Figure 5-2 shows the plots of the number of γ-rays vs. energy.

† We neglect here the small relativistic effect.

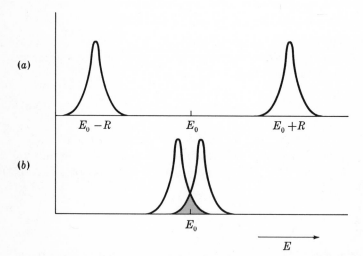

Fig. 5-2 Plots of the number of γ-rays vs. energy. (*a*) Resonance absorption and emission curves for $2R > \Gamma$. (*b*) Resonance absorption and emission curves for $2R \sim \Gamma$.

Let us now consider the Fe^{57} case which represents the typical values for a nuclear fluorescence experiment. We have $E_0 = 14.4$ keV and $M = 1.67 \times 10^{-24}$ G, and hence $R = 2 \times 10^{-3}$ eV. The half-life of the lowest excited state is 10^{-7} sec, which corresponds to a linewidth of about 5×10^{-9} eV. Obviously no resonance can be observed in this case. Earlier attempts to observe the resonance were made by supplying the source a Doppler shift. Experimentally this was done by moving the source toward the absorber with velocity v such that

$$\frac{v}{c} E_0 = 2R \qquad (5\text{-}10)$$

In this way the loss of energy due to recoil is made up, and the resonance condition is satisfied. However, although the emission and absorption lines are now brought into coincidence, this approach suffers from the serious drawback that it is not possible to restore the natural linewidth of the lines from their recoil-broadened state.

What Mössbauer discovered in 1958 was that by observing the resonance of nuclei bound in solid at low temperatures, no recoil energy was lost, and extremely sharp lines corresponding to their natural lifetime could be obtained. Because of his discovery, γ-ray fluorescence spectroscopy is now more commonly known as the Mössbauer effect. In solids, the energy of the recoiling atom can be dissipated among the lattice vibration or the solid as a whole. Since the energy of lattice vibration (also known as phonon energy) is quantized, the recoil energy can be dissipated to the solid only if it is not exactly equal to a phonon. However, because of the very large difference in mass between the recoiling atom and the solid, the energy which goes into the motion of the entire solid is extremely small and can be neglected. Consequently the emitted γ-rays will have energy E_0 if the state of the lattice remains unchanged. Similarly no energy is lost to recoil for the absorber nucleus. The usefulness of Mössbauer spectroscopy lies in the fact that because of the extremely sharp linewidth now accessible, it is possible to investigate very small changes in the nuclear energy levels which are influenced by the electronic structure of the molecules.

No Mössbauer effect can be observed in nonviscous liquids and gases because of recoil energy loss. Among the nuclei which have been found to show Mössbauer effect are Fe^{57}, Ni^{61}, Zn^{67}, Sn^{117}, I^{127}, I^{129}, Xe^{131}, Pt^{195}, and Au^{197}.

5-2 EXPERIMENTAL TECHNIQUES

The experimental setup for a Mössbauer experiment can be fairly simple (Fig. 5-3). The source is moved toward the absorber with velocity v

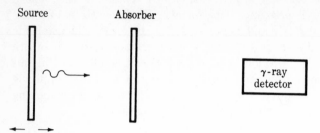

Source   Absorber

γ-ray
detector

Fig. 5-3 Block diagram of a Mössbauer spectrometer.

by means of a velocity drive which can be a turntable. The intensity
of γ-ray irradiation is measured as a function of the velocity. Figure 5-4
shows a typical experimental spectrum. The sign convention is that
positive velocity indicates the source and absorber moving toward each
other. It is important to note here that the Doppler velocity given to
the source relative to the absorber is *not* to compensate the recoil energy
loss (there is none). The Doppler velocity is necessary because of the
finite width of the resonance line and also because there are cases where
the energy level separation in the source is slightly different from that in
the absorber because of the difference in electronic configuration, as we
shall see in the next section.

Since the recoil energy of some of the emitting atoms is also shared
among lattice vibrations, the magnitude of the Mössbauer effect depends
on the recoil-free fraction f of the total emitting atoms. This fraction

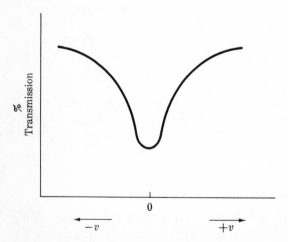

Fig. 5-4 Plot of the relative counts of γ-rays vs.
Doppler velocity.

is known as the Debye-Waller factor and is given by

$$f = \exp\left[-\frac{E_0}{k\Theta}\left(\frac{3}{2} + \frac{\pi^2 T^2}{\Theta}\right)\right] \tag{5-11}$$

where T is the absolute temperature, k is Boltzmann's constant, and Θ is the Debye temperature of the solid. It is clear, therefore, that the Mössbauer effect is greater the lower the temperature. The value of E_0 in Eq. (5-11) poses a restriction on the kind of nuclei that can be employed in such an experiment.

5-3 THEORY

There are three main hyperfine interactions that can be observed by Mössbauer spectroscopy. They are: (1) Isomer shift (δ), (2) quadrupole splittings (Δ), and (3) nuclear Zeeman splittings. These interactions and their temperature dependence can provide us with a great deal of valuable chemical information. We shall now discuss each term below.

ISOMER SHIFTS (δ)

In a Mössbauer experiment we are dealing with nuclei which are embedded in electron clouds. The electrostatic interaction between the nucleus and its surrounding electrons can shift the nuclear energy levels. The shift in energy δE is given by[1]

$$\delta E = \frac{2\pi}{5} Ze^2|\psi(0)|^2 R^2 \tag{5-12}$$

where Z is the atomic number, e the unit of charge, R the radius of the nucleus, and $|\psi(0)|$ the electronic wave function at the nucleus. Since only s electrons have nonvanishing wave functions at the nucleus, p and d electrons therefore do not cause the shift in energy levels. Applying Eq. (5-12) to the source and absorber we see immediately that if there is any difference in $|\psi(0)|$, the resonance condition will be modified as shown in Fig. 5-5. Since the radius of the nucleus in the excited state

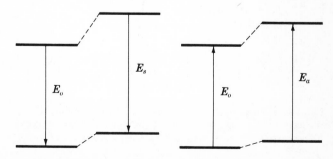

Fig. 5-5 Shifts in energy levels in the source and absorber.

R_{ex} is not equal to that in the ground state R_{gd}, the energy emitted by the source E_s is given by

$$E_s = \frac{2\pi}{5} Ze^2 |\psi_s(0)|^2 (R_{ex}{}^2 - R_{gd}{}^2) + E_0 \qquad (5\text{-}13)$$

and that for the absorption in the absorber E_a is

$$E_a = \frac{2\pi}{5} Ze^2 |\psi_a(0)|^2 (R_{ex}{}^2 - R_{gd}{}^2) + E_0 \qquad (5\text{-}14)$$

The isomer shift is given by

$$\delta = E_a - E_s = \frac{2\pi}{5} Ze^2 (|\psi_a(0)|^2 - |\psi_s(0)|^2)(R_{ex}{}^2 - R_{gd}{}^2) \qquad (5\text{-}15)$$

An isomer shift will therefore be observed if the electronic configurations around the nuclei are different in the source and the absorber. Figure 5-6 shows the isomer shift measured in ferricinium bromide. Note that the shift is of the order of 2×10^{-8} eV, which is extremely small compared with the energy of the γ-ray. However, because of the very narrow linewidth it is possible to detect such minute disturbances on the energy levels.

The isomer shift defined by Eq. (5-15) is only a relative quantity; its usefulness lies in the *comparison* of a number of compounds containing a particular Mössbauer nucleus with the same source. For example, in the study of iron compounds using Co^{57} as source it is found that δ decreases with the increasing s electron density around the nucleus, thus causing a shift of the resonance lines toward the negative velocities. This can be understood by considering the electronic configuration of iron

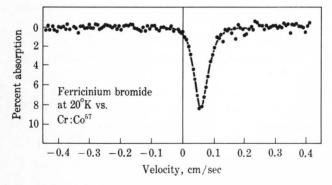

Fig. 5-6 Isomer shift in ferricinium bromide. [*From G. K. Wertheim and R. H. Herber, J. Chem. Phys.*, **38**: *2106 (1963). By permission of the authors and the American Institute of Physics.*]

in the various oxidation states

Fe (metallic) $[Ar]3d^7 4s^1$
Fe^{++} $[Ar]3d^6$
Fe^{3+} $[Ar]3d^5$

Although only s electrons contribute to $|\psi(0)|$, di- and trivalent iron salts usually have quite different δ values. The reason is that the $3s$ wave function extends farther away from the nucleus than those of $3p$ and $3d$, and hence the $3s$ electrons are screened not only by the inner cores but also by the $3p$ and $3d$ electrons. The $3s$ electrons on Fe^{++} ions are screened to a greater extent because of the additional d electron and consequently Fe^{++} salts are shifted toward more positive values compared to the Fe^{3+} salts.† The situation is less straightforward with covalent compounds since the number of electrons on iron can no longer be definitely assigned. However, a large number of iron compounds have now been measured to give a fairly satisfactory correlation between the shift and the covalency of the atom.[2]

QUADRUPOLE SPLITTINGS (Δ)

In many of the compounds studied it is found that more than one line is observed even when all the atoms in the absorber are chemically equivalent. An example of this is the Fe(CO)$_5$ compound shown in Fig. 5-7.

† This follows from the fact that R_{ex} is smaller than R_{gd} for iron, and $|\psi_s(0)|^2$ is greater than $|\psi_a(0)|^2$ if Co^{57} atoms are used as the source, as is usually the case.

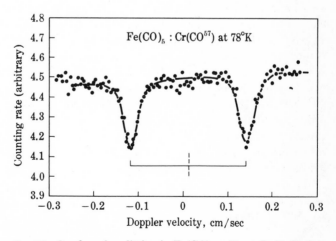

Fig. 5-7 Quadrupole splitting in Fe(CO)$_5$. [*From R. H. Herber, W. R. Kingston, and G. K. Wertheim, Inorg. Chem., **2**: 153 (1963). Copyright © 1963 by the American Chemical Society. Reprinted by permission of the copyright owner.*]

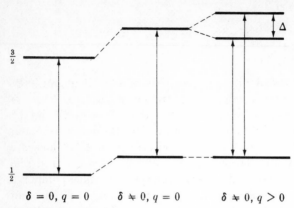

$$\delta = 0, q = 0 \qquad \delta \neq 0, q = 0 \qquad \delta \neq 0, q > 0$$

Fig. 5-8 Fe^{57} energy level splittings as a result of the quadrupole interaction.

This quadrupole splitting effect can be readily understood by considering the following energy level diagram (Fig. 5-8). If the Fe^{57} nucleus in the absorber is subject to an electric field gradient q that is not spherically symmetric, the excited-state energy level which corresponds to $I = \frac{3}{2}$ will be split into two; hence, a two-line spectrum.

NUCLEAR ZEEMAN SPLITTINGS

In addition to the quadrupole splittings, the nuclear energy levels can also be split by the magnetic field produced by the electrons. In the case of Fe^{57} both the ground and excited state will be split as shown in Fig. 5-9. A total of six lines are expected and indeed observed in some compounds. The internal magnetic field at the Fe nucleus in the metallic state has been calculated to be 3.3×10^5 G from the splittings.

Combined magnetic and quadrupole splittings are also possible although relatively few compounds show such effects. In general, in order to simplify the spectra, the source material is selected to give a single transition line. For iron this is achieved by using Co^{57} diffused into metallic chromium and no net electric field gradient and magnetic field are present at the cobalt nucleus.

5-4 APPLICATIONS

Nature of chemical bond Iron forms a large number of complexes with inorganic as well as organic ligands. In many of these compounds both π and σ bondings have been postulated between the metal atom and the ligands. Mössbauer studies have provided us with very useful informa-

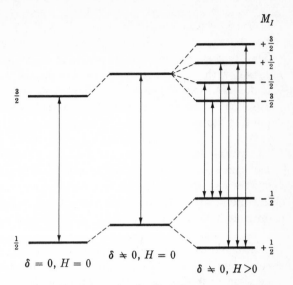

M_I

$+\frac{3}{2}$

$+\frac{1}{2}$

$-\frac{1}{2}$

$-\frac{3}{2}$

$\frac{3}{2}$

$-\frac{1}{2}$

$\frac{1}{2}$

$+\frac{1}{2}$

$\delta = 0,\, H = 0$ $\delta \neq 0,\, H = 0$

$\delta \neq 0,\, H > 0$

Fig. 5-9 Fe^{57} energy level splittings as a result of the magnetic interaction.

tion regarding the nature of these bonds. For example, in $Fe(CN)_6{}^{4-}$ both π and σ bonds are involved:

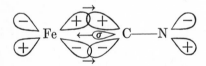

the arrows in the illustration indicate the direction of electron donation. The cyanide group donates a pair of electrons to the metal atom through the σ bond and some of the d electron density is back-donated to the ligand through d_π-p_π interaction. When one of the cyanide groups is replaced by an amine group of the type NR_3 which cannot form π bonds, a larger value of δ is observed. This follows from the fact that an increase of d electron density on iron will result in a decrease of the s electron wave function at the nucleus. In this way it is possible to list a series of ligands in the increasing order of π bond strength as follows

$$NH_3 < NO_2{}^- < Ph_3P < SO_3{}^2 < CN^- < CO < NO^+$$

The metal-ligand σ bonds apparently have little effect on the isomer-shift values in this case.

The electron withdrawing power of an electronegative substituent will of course alter the electron density at the central atom. In the study of the $Sn^{119}X_4$ compounds (where $X = F$, Cl, Br, or I), a linear correlation between δ and the electronegativity of the substituent is obtained. The data show that δ increases with increase in electronegativity consistent with the fact that Sn is sp^3 hybridized. No quadrupole splitting is observed except for SnF_4, which exists as a polymer in which the tin atom is octahedrally bonded to four bridging and two nonbridging fluorine atoms.

Structural determination   Information regarding the symmetry of the molecule can often be obtained from the quadrupole splittings.   Mössbauer studies can help us to rule out certain structures although they cannot unequivocally determine the correct one.   For example, in $Fe(CN)_6{}^{4-}$ only one line is observed for Fe^{57}, indicating a regular octahedral arrangement of the cyanide groups about Fe.   A more complicated case is that of iron dodecacarbonyl, $Fe_3(CO)_{12}$.   The Mössbauer spectrum of this compound consists of three lines as shown in Fig. 5-10.   There are three possible structures consistent with the formula as shown in Fig. 5-11. Since a maximum of two lines (in the absence of magnetic splittings) can be obtained from three equivalent iron atoms, structure (c) is immediately ruled out.   According to structures (a) and (b), only two of the three iron atoms are equivalent; these three lines are therefore due to a single-line spectrum plus a quadrupole-split two-line spectrum.   Since the central

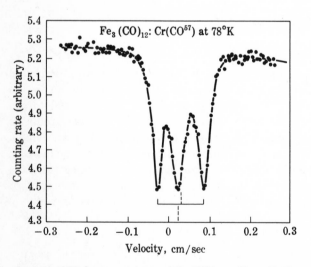

Fig. 5-10   Mössbauer spectrum of $Fe_3(CO)_{12}$.   [*From R. H. Herber, W. R. Kingston, and G. K. Wertheim, Inorg. Chem.,* **2:** *153 (1963). Copyright © 1963 by the American Chemical Society.   Reprinted by permission of the copyright owner.*]

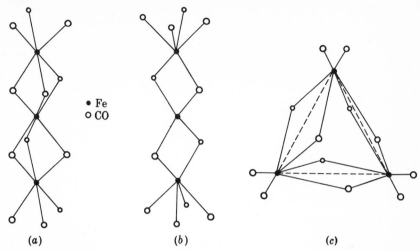

• Fe
○ CO

(a)   (b)   (c)

Fig. 5-11 Possible structures of $Fe_3(CO)_{12}$. [*From G. K. Wertheim, "Mössbauer Effect: Principles and Applications," Academic Press, Inc., New York, 1964. By permission of the Academic Press, Inc.*]

atom in (a) has a near octahedral arrangement and is expected to give rise to a single line, this structure is chosen to be the most probable one. Structure (b) would give rise to four lines due to quadrupole splittings at the two different iron atoms.

Biological applications   It is indeed fortunate that Fe^{57}, the simplest Mössbauer atom, occurs extensively in biological systems such as hemoglobin, myoglobin, cytochromes, ferritin, ferridoxin, and xanthine oxidase. Especially important is the fact that in hemoglobin, which is one of the most studied biological molecules, the iron is situated at the site of its primary activity.   The structure of heme is shown in Fig. 5-12.   The heme molecule is attached to the globin (and hence the name hemoglobin) through a nitrogen atom in an imidazole of a histidine residue. The sixth coordination site of the iron is available for formation with ligands such as H_2O, O_2, CN^-, CO, F^-, and N_3^-.   In its normal function hemoglobin contains the ferrous iron.   In the oxygenated material, that is, in oxyhemoglobin, the oxygen is assumed to lie parallel to the plane of the heme.   A comparison of the Mössbauer spectra of hemoglobin and oxyhemoglobin shows a decrease in δ for the latter compound suggesting some d-orbital π-bond interaction between iron and oxygen.   A fairly large quadrupole splitting is also observed in oxyhemoglobin due to the disturbance of the iron orbitals by oxygen.   Finally, the similar spectra

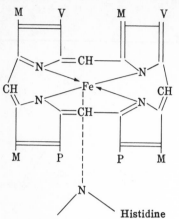

$$M = -CH_3$$
$$P = -CH_2 -CH_2 \cdot COOH$$
$$V = -CH = CH_2$$

Fig. 5-12   Central structure of heme.

of hemoglobin and hemoglobin–carbon dioxide indicates that the CO_2 molecule is not directly bonded to the iron atom.[3]

PROBLEMS

5-1. The energy loss due to recoil in nuclear fluorescence experiments must also exist in atomic fluorescence experiments.   The lifetimes associated with the electronic levels are of the order of 10^{-8} sec and the excitation energies are of the order of electron-volts.   Show that the recoil energy loss in the atomic case does not pose a similar problem in observing resonance as in the nuclear case.

5-2. Calculate the recoil energy for the 14.4-keV γ-ray decay in Fe^{57}.   If this recoil energy is not shared by the solid as a whole, what is the Doppler velocity needed to compensate the energy loss for observing the resonance?

5-3. The isomer shift δ of Fe^{57} in ferricinium bromide is 0.052 cm sec^{-1}.   Convert this value into the electronvolts.

5-4. Mössbauer resonance has also been observed with I^{127} and I^{129}.   Is it possible to deduce the nature of bonding discussed in Prob. 4-2 by measuring the isomer shift of the central iodine atom as a function of the electronegativity of X?

SUGGESTIONS FOR FURTHER READING

Introductory

Goldanskii, V. I.: "The Mössbauer Effect and Its Application to Chemistry," D. Van Nostrand Company, Inc., Princeton, N.J., 1966.

Intermediate-advanced

Frauenfelder, H.: "The Mössbauer Effect," W. A. Benjamin, Inc., New York, 1962.
Goldanskii, V. I., and Herber, R. H. (eds.): "Chemical Applications of Mössbauer Spectroscopy," Academic Press, Inc., New York, 1968.

Herber, R. H. (ed.): The Mössbauer Effect and Its Application in Chemistry, *Advan. Chem. Ser.*, **68**: 1967.
Wertheim, G. K.: "Mössbauer Effect," Academic Press, Inc., New York, 1964.

READING ASSIGNMENT

Introduction to Mössbauer Spectroscopy, R. H. Herber, *J. Chem. Educ.*, **42**: 180 (1965).

REFERENCES

1. Wertheim, p. 49.
2. L. R. Walker, G. K. Wertheim, and V. Jaccarino, *Phys. Rev. Letters*, **6**: 98 (1961).
3. G. Lang, *J. Appl. Phys.*, **38**: 915 (1967).

6

Electron Spin Resonance Spectroscopy

6-1 INTRODUCTION †

The fundamental properties of an electron are (1) mass, (2) charge, (3) spin, or intrinsic angular momentum. The spinning motion of the electron acts as a circular current and so generates a magnetic field. But the magnetic field of a circular current is equivalent to that of a magnetic dipole of moment μ_S which is given by

$$\mu_S = -g \frac{e}{2m_e c} \hbar \sqrt{S(S + 1)} \tag{6-1}$$

where g is the g factor for a free electron ($g = 2.0023$), e and m_e are the charge and mass of the electron, and $\sqrt{S(S + 1)}\, \hbar$ is the length or magnitude of the spin angular momentum vector. The negative sign indi-

---

† In principle there is no difference between NMR and electron spin resonance (ESR); both are concerned with the change in the Zeeman levels. For completeness sake we shall treat this topic along the same lines as NMR.

cates that the magnetic moment vector is in the opposite direction to that of the angular momentum vector.

When the electron is introduced into a uniform magnetic field of strength H the electron magnetic dipole will precess about the axis of the field. Analogous to the NMR case the Larmor frequency of precession ω is given by

$$\omega = \gamma H \tag{6-2}$$

where γ is the magnetogyric ratio of the dipole, that is, the ratio of the magnetic moment to the angular momentum. The magnetic energy of interaction E is given by

$$E = -\mu_S H \cos \theta \tag{6-3}$$

where θ is the angle between the axis of the dipole and the field direction. Once again because of the restriction imposed by quantum mechanics, there can be only certain values for θ. Figure 6-1 shows the vector representation of the relation between the angular momentum and its components along the axis of quantization, that is, the external magnetic field.

For an electron having spin S of $\frac{1}{2}$, there are only two possible values of θ. The projection of the spin angular momentum vector onto the axis of quantization gives $\frac{1}{2}\hbar$ and $-\frac{1}{2}\hbar$. This can be written as $M_S\hbar$, where M_S is the magnetic spin quantum number which has the values of $\pm\frac{1}{2}$. It can be easily shown that θ in this case is either $35°15'$ or $144°45'$. The same is true for the magnetic moment vector except that its direction is exactly opposite to that of the angular momentum vector (Fig. 6-2).

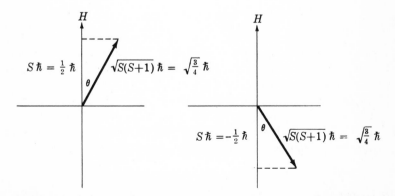

Fig. 6-1 Electron spin angular momentum vector and its components along the external magnetic field axis.

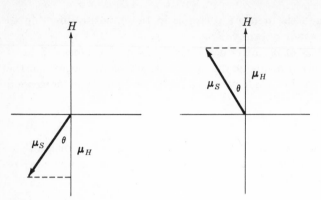

Fig. 6-2 Electron magnetic moment vector and its components along the external magnetic axis.

Next we derive the resonance condition for an unpaired electron. Let μ_H be the component of μ_S in the direction of the field so that

$$\cos \theta = \frac{\mu_H}{\mu_S} \tag{6-4}$$

Eq. (6-3) then becomes

$$E = -\mathbf{\mu}_H \cdot \mathbf{H} \tag{6-5}$$

where μ_H is given by

$$\mu_H = -g\beta M_S \tag{6-6}$$

where β is the electronic Bohr magneton, equal to $(e/2m_ec)\hbar$ (see Appendix 3). Substituting Eq. (6-6) into (6-5) we obtain

$$E = g\beta H M_S \tag{6-7}$$

For a spin of $\frac{1}{2}$, the electron has a lower energy state of $-\frac{1}{2}g\beta H$, which corresponds to the parallel alignment of the magnetic moment to the applied field, and a higher energy state of $\frac{1}{2}g\beta H$, which corresponds to the antiparallel situation. In the absence of the field the two states corresponding to $M_S = \pm\frac{1}{2}$ have the same energy and are said to be degenerate. This degeneracy is removed when the field is applied and the splitting of the levels is shown in Fig. 6-3. If electromagnetic radiation of frequency ν is present which satisfies the resonance condition

$$\Delta E = h\nu = g\beta H \tag{6-8}$$

transitions between these Zeeman levels can occur. This is the simplest case of ESR. For an assembly of unpaired electrons, the ratio of the

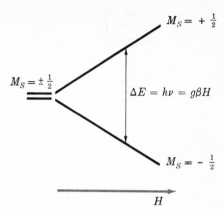

$$M_S = + \frac{1}{2}$$

$$M_S = \pm \frac{1}{2}$$

$$\Delta E = h\nu = g\beta H$$

$$M_S = -\frac{1}{2}$$

H

ig. 6-3 The splitting of the electron ceman levels in a magnetic field.

lectron population in the $M_S = \frac{1}{2}$ state, $n_{\frac{1}{2}}$, to that in the $M_S = -\frac{1}{2}$ tate, $n_{-\frac{1}{2}}$, is given by

$$\frac{n_{\frac{1}{2}}}{n_{-\frac{1}{2}}} = e^{-\Delta E/kT} = e^{-g\beta H/kT} \tag{6-9}$$

-2 EXPERIMENTAL TECHNIQUES

ESR experiments are performed by detecting the amount of energy bsorbed; therefore, in order to improve the sensitivity of detection we vant to have as many electrons in the lower state as possible. According o Eq. (6-9) this can be achieved by either reducing the temperature or ncreasing the field or both. From Eq. (6-8) we see that as H increases o does ν, and it is the frequency that turns out to have a practical upper imit. This results from the difficulty encountered in generating and letecting frequencies greater than 36,000 MHz. The most common xperimental arrangements employ frequency either in the X-band or K-band region as shown in Table 6-1. Both of the frequencies are in he microwave region, and this is why ESR is often treated as a branch of microwave spectroscopy.

Table 6-1 The frequency, wavelength, and magnetic field values for the common ESR experiments

Band	ν (MHz)	λ (cm)	H (G)
X	9,500	3	3,400
K	36,000	0.8	13,000

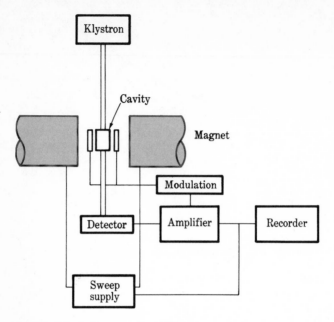

Fig. 6-4 Block diagram of an ESR spectrometer.

The basic features of an ESR spectrometer are (1) a source of micro-wave radiation, (2) a sample cell, (3) a means of transmitting the radia-tion energy to the sample cell, (4) a dc magnetic field, (5) a detection system, and (6) a recorder or an oscilloscope (Fig. 6-4). The usual source of radiation is a klystron oscillator which produces monochromatic microwave radiation. This radiation is transmitted into the sample cell by means of a waveguide, which is a copper or brass tubing of dimensions appropriate to the wavelength of radiation. The sample cell is called the cavity and is located in a homogeneous magnetic field. The trans-mitted radiation is detected in a rectifier crystal and the signal, after amplification, is either displayed on an oscilloscope or recorded perma-nently on the chart paper. Instead of displaying the signal as the absorption curve as with NMR experiments, ESR signals are usually *phase-sensitive* detected and represented as the first derivatives.[1] The absorption curve and the first and second derivatives of an ESR line are shown in Fig. 6-5.

6-3 THEORY

In Sec. 6-1 we saw the simplest example of an ESR experiment, that is, the resonance of a free electron. Isolated electrons are of little interest and we are naturally more interested in situations in which we can use the electron as a probe to obtain useful information about the atoms and molecules. In most molecules electrons regularly occur in pairs with opposite spins as required by the Pauli exclusion principle; hence ESR does not have as wide an application as NMR. The same restriction,

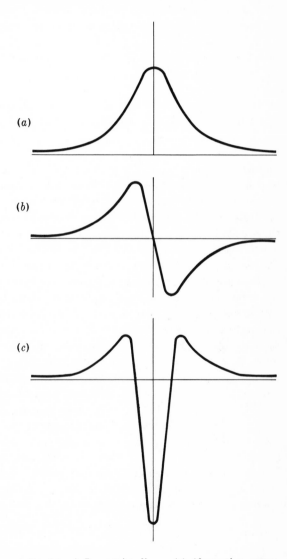

Fig. 6-5 A Lorentzian line. (*a*) Absorption curve. (*b*) First-derivative curve. (*c*) Second-derivative curve.

however, also makes it a unique tool in the study of paramagnetic systems. The following is a list of systems containing unpaired electrons.

1. Conduction electrons of metals
2. Semiconductors
3. Transition metal ions
4. Molecules containing odd number of electrons (NO, NO_2, ClO_2, and NF_2)
5. Radicals produced by chemical or physical means
6. Triplet state molecules

In this chapter we shall discuss only the ESR aspects of 5 and 6.

The next simplest case of ESR is that of a hydrogen atom. The proton has a nuclear spin I of $\frac{1}{2}$ with a resultant magnetic moment which can be aligned either parallel or antiparallel to the electron magnetic moment. This interaction further splits the electron Zeeman levels into four levels and additional transitions are now possible. The transitions are restricted by the selection rules $\Delta M_S = \pm 1$ and $\Delta M_I = 0$, so only two lines can be detected. These two lines are of equal intensity, and the separation between them gives the hyperfine splitting constant (HSC), which is a measure of the extent of this electron-nucleus interaction. The additional structure (lines) is called the hyperfine structure. When two protons are present which interact equally with the electron there will be six levels and three allowed transitions (Fig. 6-6). In general, for n equivalent protons† there are $n + 1$ lines; the relative intensities of the lines are proportional to the coefficients of a binomial expansion of order n, that is, $(1 + x)^n$. In Fig. 6-6 resonance is achieved by varying the microwave frequency (constant magnetic field). In practice, however, because of practical convenience, ESR experiments, like NMR, are always performed by varying the field (constant frequency).

In addition to the intrinsic spin angular momentum discussed in Sec. 6-1, an electron may also possess an orbital angular momentum in atoms and molecules. The magnitude of the orbital angular momentum is given by $\sqrt{L(L + 1)}\,\hbar$, where L is the azimuthal quantum number. There will then be two contributions to the electron magnetic moment, one due to the spin and the other due to orbital motion. The resultant magnetic moment μ_J is given by

$$\mu_J = -g_J\beta \sqrt{J(J + 1)}\,\hbar \qquad (6\text{-}10)$$

where J is the resultant angular momentum which in the Russell-Saunders

† In ESR, equivalent protons are those that have the same HSCs as a result of symmetry.

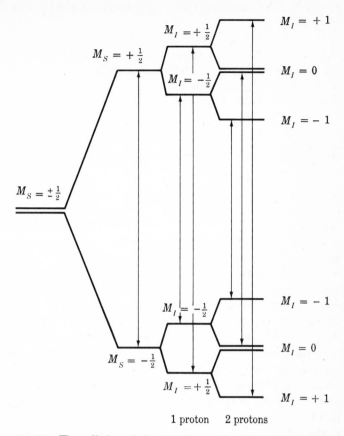

1 proton   2 protons

Fig. 6-6 The splitting of electron Zeeman levels as a result of the interaction with one and two protons. (Resonance is achieved by fixing the magnetic field and varying the frequency in this case.)

coupling scheme is given by (see Chap. 10)

$$\mathbf{J} = \mathbf{L} + \mathbf{S} \tag{6-11}$$

and g_J, the Landé g factor (see Appendix 6) is given by

$$g_J = 1 + \frac{J(J+1) + S(S+1) - L(L+1)}{2J(J+1)} \tag{6-12}$$

Thus g_J may be appreciably different from 2.0023 if there is coupling between the orbital and the spin angular momentum. However, for most organic free radicals, because of the high asymmetry of the molecules, the orbital angular momentum is not conserved, and consequently we say that L is not a "good" quantum number. Hence we have $L \simeq 0$

and $J \simeq S$. Therefore, their g values lie very close to that of the free electron.

ESR theory is simplest for liquid solutions and it is in these systems that most of its chemical applications are found. Let us now consider the various magnetic interactions of organic radicals containing nuclei with spin I. In the presence of an external magnetic field the various magnetic interactions are represented by the hamiltonian $\mathfrak{K}$ where

$$\mathfrak{K} = \mathfrak{K}_{\text{Zeeman}} + \mathfrak{K}_{\text{dip}} + \mathfrak{K}_{\text{sc}} \tag{6-13}$$

where $\mathfrak{K}_{\text{Zeeman}}$ represents the interaction of the electron and the nucleus with the applied field

$$\mathfrak{K}_{\text{Zeeman}} = -g_N \beta_N \mathbf{I} \cdot \mathbf{H} + g\beta \mathbf{S} \cdot \mathbf{H} \tag{6-14}$$

$\mathfrak{K}_{\text{dip}}$ represents the classical dipole-dipole interaction between the electron and nuclear magnetic moments

$$\mathfrak{K}_{\text{dip}} = gg_N \beta\beta_N \left[\frac{\mathbf{S} \cdot \mathbf{I}}{|\mathbf{r}_e - \mathbf{r}_n|^3} - \frac{3(\mathbf{S} \cdot \mathbf{r})(\mathbf{I} \cdot \mathbf{r})}{|\mathbf{r}_e - \mathbf{r}_n|^5} \right] \tag{6-15}$$

where $\mathbf{r}_e$ is the vector position of the electron and $\mathbf{r}_n$ is the vector position of the nucleus. In nonviscous liquids this interaction vanishes because of rapid tumbling of the molecules just as in the NMR case. Of course the same type of interaction also exists between unpaired electrons on different molecules, and we might expect that this effect would be much larger because of the much larger magnetic moment of the electron. However, if we study only very dilute solutions of these radicals, which is often the case, such electron-electron dipolar interaction can be neglected.

$\mathfrak{K}_{\text{sc}}$, which is also called the Fermi contact interaction, depends only on the electron "contact" with the nucleus. It has no classical counterpart and does not vanish in solution. We have[1a]

$$\mathfrak{K}_{\text{sc}} = gg_N \beta\beta_N \frac{8\pi}{3} \mathbf{S} \cdot \mathbf{I}\delta(\mathbf{r}_e - \mathbf{r}_n) \tag{6-16}$$

The most important term in Eq. (6-16) is the Dirac delta function $\delta(\mathbf{r}_e - \mathbf{r}_n)$, which will have a nonzero value only if the wave function of the unpaired electron in the radical has a nonvanishing value at the nucleus.

The total energy of various magnetic interactions is given by

$$E = \int \psi \mathfrak{K}\psi \, d\tau \tag{6-17}$$

where ψ contains both the space and the spin part of the wave function of the electron. The spin part of the wave function, which is denoted by α or β, gives

$$\int \alpha(g\beta\mathbf{S}\cdot\mathbf{H})\alpha\,d\tau = \tfrac{1}{2}g\beta H \qquad \int \beta(g\beta\mathbf{S}\cdot\mathbf{H})\beta\,d\tau = -\tfrac{1}{2}g\beta H$$

Thus in general we write

$$\int M_S(g\beta\mathbf{S}\cdot\mathbf{H})M_S\,d\tau = g\beta H M_S$$

If we consider the situation in solution then $\mathcal{3C}_{\text{dip}}$ vanishes. The nuclear Zeeman interaction is much smaller than that for the electron and is also neglected. Equation (6-17), when written in the Paschen-Back limit (see Chap. 10), becomes†

$$E = \int \psi \left\{ g\beta\mathbf{S}\cdot\mathbf{H} + gg_N\beta\beta_N \frac{8\pi}{3} \mathbf{S}\cdot\mathbf{I}\delta(\mathbf{r}_e - \mathbf{r}_n) \right\} \psi\,d\tau$$

$$= g\beta H M_S + a M_I M_S \tag{6-18}$$

a is the HSC mentioned earlier and is given by

$$a = \frac{8\pi}{3}\, gg_N\beta\beta_N|\psi(0)|^2 \tag{6-19}$$

Since $\psi(0)$ is the value of ψ at the nucleus, we would expect electron-nucleus contact interaction only if ψ has a finite, nonzero value at the nucleus (see Fig. 1-6). It follows, therefore, that in order to have this contact interaction ψ must have some s character.

Applying Eq. (6-18) to the hydrogen atom we obtain

$$M_S = \tfrac{1}{2} \begin{cases} E_1 = \tfrac{1}{2}g\beta H + a(\tfrac{1}{2})(\tfrac{1}{2}) \\ E_2 = \tfrac{1}{2}g\beta H + a(-\tfrac{1}{2})(\tfrac{1}{2}) \end{cases}$$

$$M_S = -\tfrac{1}{2} \begin{cases} E_3 = -\tfrac{1}{2}g\beta H + a(-\tfrac{1}{2})(-\tfrac{1}{2}) \\ E_4 = -\tfrac{1}{2}g\beta H + a(\tfrac{1}{2})(-\tfrac{1}{2}) \end{cases} \tag{6-20}$$

The two allowed transitions are

$$E_1 - E_4 = g\beta H + \frac{a}{2}$$
$$E_2 - E_3 = g\beta H - \frac{a}{2} \tag{6-21}$$

The two resonance lines are shown in Fig. 6-7. When more than one type of proton is present, Eq. (6-18) can be written in the general form

$$E = g\beta H M_S + a_1 M_S \sum_{i=1}^{n} M_I{}^i + a_2 M_S \sum_{i=1}^{n} M_I{}^i + \cdots \tag{6-22}$$

where $a_1, a_2, \ldots$ are the HSCs of proton types $1, 2, \ldots$, and n is the number of protons for a particular type. Experimentally, HSCs are

† Both the space and spin part of ψ are used to evaluate the second term in Eq. (6-18).

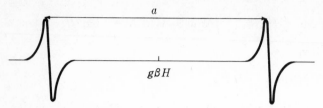

Fig. 6-7   First-derivative ESR spectrum of hydrogen atom.

usually measured in gauss although they can also be expressed in megahertz.   The conversion between gauss and megahertz is as follows:

$$a \text{ (G)} = 0.35683 \left(\frac{g_{\text{free electron}}}{g_{\text{radical}}} \right) a \text{ (MHz)}$$

The ESR spectrum of the hydrogen atom produced by electric discharge or trapped in solids shows the expected two-line spectrum with a separation of 506 G.   This is the largest splitting for a proton since the unpaired electron is in the $1s$ orbital and so has 100 percent s character.

Next, let us consider a simple organic free radical, the benzene anion radical, prepared by the chemical reduction of benzene with alkali metal in an inert solvent such as 1,2-dimethoxyethane or tetrahydrofuran.   A spectrum containing seven lines with relative intensities 1:6:15:20:15:6:1 and a HSC of 3.75 G is obtained (Fig. 6-8).   In the simple molecular orbital approximation, the unpaired electron occupies the antibonding molecular orbital, which has a node in the plane of the molecule, that is, the wave function is zero at the nucleus.   How do we then account for

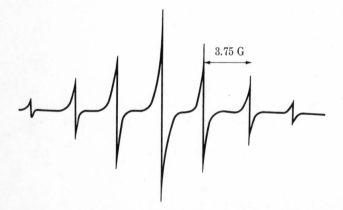

Fig. 6-8   First-derivative ESR spectrum of benzene anion radical.

the observed hyperfine structure? Qualitatively, the origin of this hyperfine interaction can be explained by considering the $> \overset{\cdot}{C}$—H fragment of an aromatic system. A C—H σ-bonding orbital is formed by the overlap of the carbon sp^2 hybrid orbital with the hydrogen $1s$ orbital and is occupied by two electrons. The unpaired electron occupies the carbon $2p_\pi$ orbital. According to the particular manner of electron pairing in this orbital, two structures are possible, as shown in Fig. 6-9. However, the structure shown in Fig. 6-9a is slightly favored over that in Fig. 6-9b because the electrons on the carbon atom have their spins parallel to each other so that there will be an *exchange interaction* between them which stabilizes the system. The result is that there will be a small amount of electron unpairing in the C—H σ-bonding orbital. This *polarization* of the electrons will produce some unpaired spin density (to be defined later) on the hydrogen atom, which accounts for the hyperfine structure observed. Of course this is only a very small effect as evidenced by the small HSC of 3.75 G (cf. HSC in hydrogen atom).

Quantitative calculations of this effect involve the use of configuration interaction and perturbation theory. A useful result is given by McConnell[2]

$$a_H = Q_{CH}\rho \qquad\qquad (6\text{-}23)$$

where a_H is the proton HSC, ρ is the unpaired spin density on the carbon atom to which the proton is attached, and Q_{CH}, which is called the spin-polarization parameter, is roughly a constant for π radicals and is usually given the value of 25 G. The unpaired spin density is defined in terms

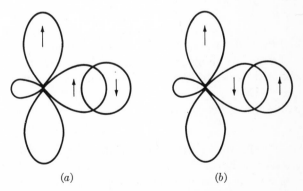

(a)   (b)

Fig. 6-9 Two possible arrangements of electron pairing in $> \overset{\cdot}{C}$H. The electrons in the carbon $2p_\pi$ and sp^2 orbitals have their spins parallel in (a) and antiparallel in (b).

of molecular orbital theory as

$$\rho_i = c_i^2 \qquad (6\text{-}24)$$

where ρ_i is the unpaired spin density on the ith carbon atom and c_i is the coefficient of the atomic orbital of the same atom in the molecular orbital containing the unpaired electron.

The usefulness of Eq. (6-23) is that a_H is an experimentally measured quantity, Q_{CH} is roughly a constant and ρ_i can be obtained for comparison with the value calculated either by molecular orbital or valence bond theory.

The spin polarization explanation that accounts for the proton hyperfine structure also predicts unpaired spin density on the carbon atom. However, C^{12}, the most abundant of the carbon isotopes (98.9 percent), does not have a nuclear spin. C^{13}, on the other hand, has a nuclear spin I of $\frac{1}{2}$ so further hyperfine structure will arise. Due to its low abundance these hyperfine lines are usually very weak, but in some favorable cases, for example, the benzene anion radical, where there is a large number of equivalent nuclei, these fine structures have indeed been observed. The situation for radicals containing nitrogen is quite different. The most abundant nitrogen isotope is N^{14}, which has a spin I of 1, and there are three quantized orientations of the nuclear spin angular momentum characterized by $M_I = 1$, 0, and -1. Interaction of an unpaired electron with one nitrogen nucleus gives rise to three equally spaced, equally intense lines, and the interaction with two equivalent nuclei will produce five lines with relative intensities $1:2:3:2:1$. N^{15} has a nuclear spin I of $\frac{1}{2}$, but because of its low abundance (0.37 percent) its hyperfine lines are not easily detected. Deuterium has a spin I of 1 and although its natural abundance is too low to be detected in any radicals (1.16×10^{-2} percent), many compounds can be synthesized with deuterium in place of hydrogen atoms. Because of its larger mass, the deuterium magnetic moment μ_D is smaller than that of the proton μ_H and the ratio of deuteron to proton HSC is given by

$$\frac{a_D}{a_H} = \frac{\mu_D I_H}{\mu_H I_D} = 0.1535 \qquad (6\text{-}25)$$

One of the most important applications of ESR to chemistry is the study of hyperfine interactions and the extent of electron delocalization in radicals. As mentioned earlier, the procedure is to compare the spin density value on atoms obtained experimentally using Eq. (6-23) with those calculated by different valence theories, for example, molecular orbital or valence bond method. It therefore provides us a convenient way to test the soundness of the various theoretical models. However, such a comparison is possible only if we can analyze the experimental spectrum, that is, if we can first determine the HSCs and then assign them to the particular atoms in the radical. The analysis of the benzene anion radical is exceedingly simple because there is only one kind of proton present and therefore only one HSC. A slightly more complex spectrum is that of the naphthalene anion radical, which is also prepared by the reduction with alkali metal in ethereal solvents. Naphthalene has two kinds of protons (four each) and therefore $(4 + 1)(4 + 1) = 25$ lines. The usual procedure of analysis is to compare the theoretical spectrum constructed using the HSCs measured together with the number of equivalent protons (or other nuclei) in the radical with the experimental spectrum. The analysis can be quite complex for radicals containing a large number of inequivalent magnetic nuclei. There is really no short-cut to this, and the process is mainly one of trial and error. Sometimes deuterium substitution can be of great aid in the analysis as well as the assignment of the spin densities to the various atoms.

Analogous to the NMR case, the shape of an ESR line can also be described by either the Gaussian or the Lorentzian function. Further-more, the discussion of the relaxation times in Appendix 4 applies equally well to the ESR case.

The foregoing discussion was focused on systems with only one unpaired electron spin, that is, the doublet state ($S = -\frac{1}{2}$ and $2S + 1 = 2$). We shall now examine the situation in which there are two unpaired spins, that is, the triplet state ($S = 1$ and $2S + 1 = 3$). An example of such a system is the ground-state oxygen molecule. There are also a number of transition metal ions which are triplets or even higher multiplets due to their incompletely filled inner d shells. Although there has been rather extensive ESR investigation on these inorganic systems we shall not be concerned with them in this chapter. Instead, we shall discuss only the photoexcited triplet molecules.

It was originally suggested by Lewis and Kasha[3] that the upper state involved in the phosphorescence of many organic molecules in rigid solvents is the lowest triplet state into which the molecule falls after being optically excited to a higher singlet state (see Chap. 12). The lifetimes of the triplet state are fairly long (of the order of seconds), the reason being that the transition from the triplet to the singlet state is *spin-symmetry-forbidden*. (The selection rule is $\Delta S = 0$.) These triplet molecules are paramagnetic, and if a high enough concentration is maintained, we might expect that ESR experiments can be performed. The hamiltonian operator for the triplet molecule can be written in a similar way to that for the doublet state:

$$\mathcal{H} = \mathcal{H}_{\text{Zeeman}} + \mathcal{H}_{\text{dip}} + \mathcal{H}_{\text{sc}} \tag{6-26}$$

where

$$\mathcal{H}_{\text{Zeeman}} = -g_N\beta_N\mathbf{I}\cdot\mathbf{H} + g\beta(\mathbf{S}_1 + \mathbf{S}_2)\cdot\mathbf{H} \qquad (6\text{-}27)$$

$$\mathcal{H}_{\text{sc}} = \frac{8\pi}{3}\,g\beta g_N\beta_N(\mathbf{S}_1 + \mathbf{S}_2)\cdot\mathbf{I}\delta(\mathbf{r}_e - \mathbf{r}_n) \qquad (6\text{-}28)$$

and

$$\mathcal{H}_{\text{dip}} = g^2\beta^2\left[\frac{\mathbf{S}_1\cdot\mathbf{S}_2}{r^3} - \frac{3(\mathbf{S}_1\cdot\mathbf{r})(\mathbf{S}_2\cdot\mathbf{r})}{r^5}\right] \qquad (6\text{-}29)$$

where $\mathbf{S}_1$ and $\mathbf{S}_2$ are the spin angular momentum vectors of electrons 1 and 2 and $\mathbf{r}$ is the vector joining them.

It might be thought that from the previous discussion on the doublet-state case that the three states corresponding to $M_S = 1, 0,$ and -1 will be degenerate in the absence of an external magnetic field. However, because of the strong coupling between the two spins the degeneracy will be removed even without the field present and the splitting of the levels is called the zero-field splitting. The size of the zero-field splitting varies according to the type of molecules, but are usually of the same order as the splitting due to a strong magnetic field (about 3000 G). Figure 6-10 shows a typical splitting pattern for the triplet state with field along one axis of spin quantization of the molecule. In a molecule without spherical symmetry, the splitting with H will vary with θ, which is the angle between the applied magnetic field and the molecular axis.

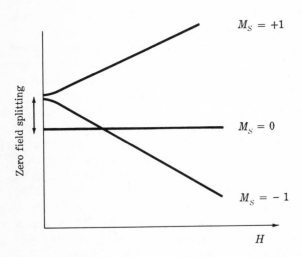

Fig. 6-10 Splitting of the electron Zeeman levels of a triplet aromatic molecule (for example, triphenylene). The external field is perpendicular to the plane of the molecule.

For a rigid solution containing randomly oriented triplet molecules, θ can have any value, and the absorption spectrum will then be a series of overlapping lines (arising from the $\Delta M_S = \pm 1$ transitions) of varying separations; the overall signal will then be too broad to be detected. This *anisotropy* effect, that is, the variation of splitting with θ, can be removed if we can somehow align the molecules in a regular manner so that there will be only one θ, and no broadening due to the overlapping of lines will result. The first successful experiment on the triplet state was performed by Hutchinson and Mangum,[4] who studied the phosphorescence state of naphthalene in a single crystal of durene at 77°K. Naphthalene and durene have similar geometry and it is possible to replace some of the durene molecules with naphthalene by forming a solid solution. All the naphthalene molecules are aligned in exactly the same manner, and there is only one θ present. By suitable rotation of the crystal with respect to the magnetic field, that is, varying θ, the resonance lines can be shifted up and down field over a range of 2100 G. This means that for randomly oriented triplet naphthalene molecules the signal will have a width of about 2100 G and will therefore be very difficult to detect. In addition to the $\Delta M_S = \pm 1$ transitions shown in Fig. 6-10, one can also observe resonance lines due to the $\Delta M_S = \pm 2$ transitions. These transitions occur at about one-half the magnetic field to the $\Delta M_S = \pm 1$ transitions and are therefore also known as the half-field resonance. The most important feature of the half-field resonance lines is that the anisotropic line-broadening effect is very much reduced, and it is often possible to study these transitions in randomly oriented molecules.[5] This is of great practical importance since one is no longer restricted by the single-crystal technique described for the naphthalene case.

Very few ESR studies of triplet molecules have been carried out in solution because (1) their lifetimes are usually too short at high temperatures and (2) the anisotropic line-broadening effect is too great.

Quantitative treatment of the triplet energy levels is beyond the scope of this book. Here we shall only note that the only important terms in Eq. (6-26) are the electron Zeeman and electron dipole-dipole interactions. The energy due to the electron-nucleus contact interaction, which is of primary importance in the doublet state in solution, is at least an order of magnitude smaller than that for the dipolar interaction. Therefore, ESR spectra of triplet molecules rarely show any resolved hyperfine structure due to this contact interaction.

Finally, we note that not all the organic triplet molecules are in the photoexcited state. For example, it had been postulated in the study of organic reaction mechanisms that certain carbenes and nitrenes react as though they were in the triplet state. Carbenes and nitrenes are usually produced by the photolysis of the particular diazo compound at 77°K:

$$C_6H_5—\underset{\substack{\|\\N_2}}{C}—C_6H_5 \xrightarrow{h\nu} \underset{C_6H_5}{\overset{C_6H_5}{\diagdown}} C: + N_2$$

$$C_6H_5—N_3 \xrightarrow{h\nu} C_6H_5—N: + N_2$$

ESR studies of these compounds have indeed confirmed the ground state triplet properties.[5]   For many of the systems studied it is found that the two unpaired spins are mainly localized on the carbon (and nitrogen) atom.

6-4 APPLICATIONS

Structural determination   The ESR technique cannot be applied to elucidate molecular structure as NMR can because the information obtained from the hyperfine structure is mostly about the extent of electron delocalization and Fermi contact interaction.   It does not tell us, for example, the arrangement of the atoms in the molecule although the symmetry of the molecule can sometimes be deduced from the sets of equivalent nuclei.   In some cases, however, ESR is able to provide useful information about the shape of the radicals.   An interesting example of this is the determination of the methyl radical structure.   The methyl radical can be produced in the solid state by high-energy irradiation of CH_3I at low temperatures, and is stabilized in the CH_3I matrix.   It may have one of the following two structures: planar and tetrahedral.   The first corresponds to a sp^2 and the second corresponds to a sp^3 hybridized carbon atom.   The ESR spectrum of $\cdot CH_3$ shows the expected four lines (from the three equivalent protons) with a splitting of 25 G.   Recalling McConnell's equation we write

$$a_H = Q_{CH}\rho$$

which is derived for planar π radicals.   Using $a_H = 25$ G and $\rho = 1$ we obtain a value of 25 G for Q_{CH}, which is very close to the value obtained for most aromatic radicals.   This would seem to suggest that methyl radical has a planar structure.   However, the observed splittings are not positive proof of the planarity of $\cdot CH_3$ for we do not know on theoretical grounds just how large the proton HSC would be if $\cdot CH_3$ were not planar. On the other hand, we can make reasonable theoretical estimates of the C^{13} HSCs in planar and tetrahedral $\cdot CH_3$.   From the ESR spectrum of the $C^{13}H_3$ radical, a splitting of 41 G is obtained.[6]   If the methyl radical were tetrahedral, the unpaired electron would be in one of the sp^3 hybrid orbitals. This would have 25 percent s character,† and the interaction with the C^{13} nucleus, according to theoretical calculations, would give a splitting of 300 G. The much smaller splitting constant observed can only be taken to mean that the unpaired electron has a very small amount of s character; that is, it is in a predominantly p_π-type orbital. Therefore, methyl radical is most likely to be planar.

Study of unstable paramagnetic species Many chemical reactions are known to involve the formation of paramagnetic intermediates at one stage or another, and their identification is important in mechanistic studies. Most of these radicals are unstable because of their high reactivities. In order to maintain a high enough steady concentration for ESR studies, the rapid-flow system is usually employed. The compounds whose interaction yields the paramagnetic intermediates are mixed shortly before the mixture flows through the cavity of the ESR spectrometer. The usual arrangement is to pass two jets of solutions into a mixing chamber which is directly under the cavity, and the mixed solution then flows through the cavity under hydraulic pressure. In this way a steady concentration of the radicals can be maintained. With proper adjustment of pressure, radicals of lifetimes of about 0.01 sec have been studied. Figure 6-11 shows the ESR spectrum of $\cdot CH_2$—OH and CH_3—$\overset{\cdot}{C}H$—OH

† The four sp^3 hybrid orbitals, in cartesian coordinates, are

$$\Phi_1 = \tfrac{1}{2}(s + p_x - p_y + p_z)$$
$$\Phi_2 = \tfrac{1}{2}(s - p_x + p_y + p_z)$$
$$\Phi_3 = \tfrac{1}{2}(s - p_x - p_y - p_z)$$
$$\Phi_4 = \tfrac{1}{2}(s + p_x + p_y - p_z)$$

where Φ_1, Φ_2, . . . are the hybrid orbitals; s, p_x, p_y, and p_z are the carbon $2s$ and $2p$ atomic orbitals.

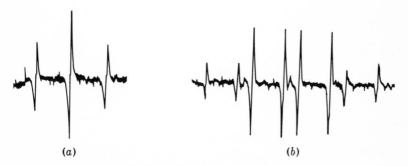

(a)   (b)

Fig. 6-11 (a) First-derivative spectrum of $\overset{\cdot}{C}H_2$—OH. (b) First-derivative spectrum of CH_3—$\overset{\cdot}{C}H$—OH. [*From W. T. Dixon and R. O. C. Norman, J. Chem. Soc., 3119 (1963). By permission of the Chemical Society and Dr. Norman.*]

using the flow technique.   These radicals were generated by mixing an acidified solution containing titanous ions ($TiCl_3$) with an acidified solution containing CH_3OH or C_2H_5OH and hydrogen peroxide.   The following reactions are known to take place[7]

$$Ti^{3+} + H_2O_2 \rightarrow Ti^{4+} + \cdot OH + OH^-$$
$$Ti^{3+} + \cdot OH \rightarrow Ti^{4+} + OH^-$$
$$\cdot OH + H_2O_2 \rightarrow H_2O + \cdot O_2H$$
$$CH_3OH + \cdot OH \rightarrow \cdot CH_2\!-\!OH + H_2O$$
$$C_2H_5OH + \cdot OH \rightarrow CH_3\!-\!\overset{\cdot}{C}H\!-\!OH + H_2O$$

Figure 6-12 shows the flow system.   Apparently only the hydroxyl radical is responsible for the abstraction of the hydrogen atom because of its greater reactivity.   With this technique a large number of unstable aliphatic and aromatic radicals are now accessible to us, and the scope of ESR solution studies is very much widened.

Kinetic studies of electron transfer reactions   Consider the situation in which a free anion radical N^- is in the presence of its neutral molecules N in solution.   The unpaired electron is no longer confined to one particular site (molecule) since there are now other equally suitable molecules to accommodate it, and the following reaction will take place:

$$N^- + N \rightleftharpoons N + N^-$$

This process involves only the transfer of an electron and can therefore

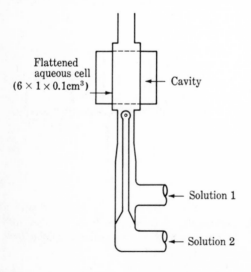

Flattened
aqueous cell
($6 \times 1 \times 0.1\,\text{cm}^3$)

Cavity

Solution 1

Solution 2

Fig. 6-12 Rapid-flow apparatus for ESR studies. [*From W. T. Dixon and R. O. C. Norman, J. Chem. Soc., 3119 (1963). By permission of the Chemical Society and Dr. Norman.*]

be thought of as an oxidation-reduction reaction. Under suitable conditions, the rate of electron jumping from one molecule to another, or the rate constant for the bimolecular reaction can be deduced from the ESR spectrum of N⁻. As an example let us consider the electron transfer reaction between a tetracyanoethylene (TCNE) anion radical and its neutral molecule. The TCNE anion radical is formed by the reduction with alkali metal in ethereal solvents, and the transfer reaction is as follows:

Figure 6-13 shows the sequence of line broadening as TCNE is added to a solution containing the TCNE anion radical. First, in the absence of TCNE, a sharp nine-line spectrum due to four equivalent nitrogen nuclei is obtained. As TCNE is added to the solution the lines will begin to broaden and then coalesce into a broad spectrum, and finally, at very high concentrations of TCNE, only one sharp line is obtained. The initial line broadening as TCNE is added is a direct consequence of Heisenberg uncertainty principle. As the lifetime of the unpaired electron at a particular molecule is shortened by the transfer process, the uncertainty in its energy becomes greater because of the relationship $\Delta E \, \Delta t \sim \hbar$. At higher concentrations of TCNE each line becomes so broad that it will overlap with the adjacent lines. At very high concentrations the electron jumps from one molecule to another so fast that all the detailed hyperfine structure is lost, and only one sharp line representing this average effect is observed. This is known as the *exchange-narrowing phenomenon*. The bimolecular rate constant can be obtained from all stages of electron transfer, but here we shall discuss only the *slow-limit case*. If the concentration of TCNE added is low so that the line broadening due to electron transfer is small compared to the separation between the lines, the bimolecular rate constant k is given by[8]

$$k = \frac{1.5 \times 10^7 (\Delta H - \Delta H_0)}{[N]}$$

where ΔH_0 and ΔH are the first-derivative linewidths in gauss of a par-

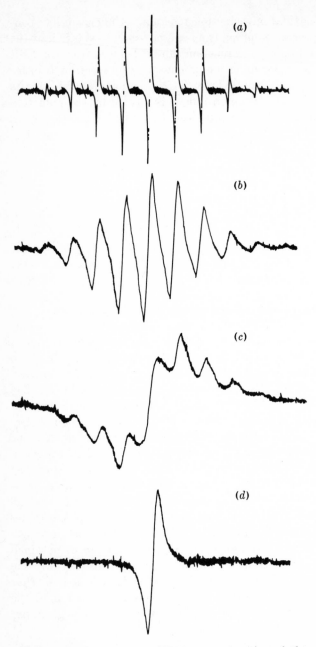

Fig. 6-13 ESR spectra of TCNE$^-$ as a function of the transfer rate. (*a*) No neutral TCNE is present. The concentration of TCNE increases from (*b*) to (*d*).

ticular line in the absence and presence of electron transfer and [N] is the concentration of the neutral compound added (in this case, TCNE). It is not required to know the concentration of the radical as long as its concentration is kept low.

The use of ESR in the study of electron transfer reactions was first reported by Ward and Weissman for the naphthalene-naphthalene anion radical system.[8] The bimolecular rate constants measured usually lie in the range of 10^7 to 10^9 l mole^{-1} sec^{-1}. Finally, we note that ESR study of electron transfer reactions is based on exactly the same principle as the NMR study of proton transfer reactions discussed in Chap. 3.

Spin-labeling studies of biomolecules Certain important information regarding the chemical, structural, and kinetic properties of biomolecules can be obtained from the ESR studies of synthetic organic free radicals which are chemically attached to these molecules. This technique makes use of the fact that by following the changes of the ESR spectrum of the radical in the free and bound state we can deduce information about the environment close to the binding site or about the macromolecule as a whole. Spin-label radicals are generally nitroxide radicals of the type

They are very stable and inert and show sharp, well-resolved spectra that are sensitive to the molecular environment. The unpaired electron is largely localized on the nitrogen atom so that its spectra usually consist of only three lines of equal intensity. In the conformational study of bovine serum albumin (BSA), a protein molecule, McConnell and Griffith prepared the following spin-label radical: N-(1-oxyl-2,2,5,5-tetramethylpyrrolidinyl)-maleimide[9]

and attached it to BSA.   The reason for choosing this particular radical
is that it is known that N-ethyl maleimide reacts specifically with the

$$
\begin{array}{c}
\text{H} \\
\diagdown \\
\text{C} \\
\parallel \\
\text{C} \\
\diagup \\
\text{H}
\end{array}
\begin{array}{c}
\text{O} \\
\parallel \\
\text{C} \\
\diagdown \\
\qquad\qquad \text{N—Et} \\
\diagup \\
\text{C} \\
\parallel \\
\text{O}
\end{array}
$$

—SH group in BSA so in order to study the environment close to the
—SH site we need a radical which contains a functional group to react
with the sulfur atom.   This radical can also be bound to an amino site
in the lysine group in BSA.   Figure 6-14 shows the ESR spectra of the
nitroxide radical when it is attached to the —SH group and when it is
attached to the lysine group.   The second spectrum is obtained from the

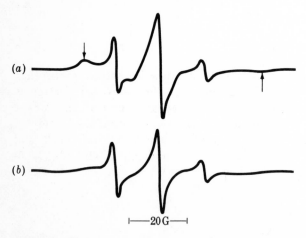

Fig. 6-14   ESR spectrum of bovine serum albumin spin-
labeled with the nitroxide maleimide.   Labeling procedures
for samples yielding (a) and (b) were identical, except that
in (b) the —SH group of the serum albumin was blocked by
reaction with N-ethyl maleimide before spin labeling.
[*From O. H. Griffith and H. M. McConnell, Proc. Nat. Acad.
Sci. U.S.*, **55:** *8 (1966).   By permission of the National Academy of Sciences and the authors.*]

compound prepared by first treating BSA with N-ethyl maleimide and then reacting the product with the nitroxide radical. This ensures the "blocking" of all the $-SH$ sites by N-ethyl maleimide so the radical can react only with the lysine group. The arrows in Fig. 6-14a indicate that the spectrum is really a superposition of two spectra of the radicals due to two different environments. The central spectrum is similar to that shown in Fig. 6-14b and suggests relative free motion. The broad-line spectrum corresponds to the situation in which the radical is immobilized; that is, it cannot rotate freely. From the line separations McConnell and Griffith calculated that the rotational time for the radical is at least 10^{-8} sec. At a pH of 2.1, where BSA is known to undergo reversible denaturation, the intensity of the signal due to the immobilized spin label is almost zero whereas the intensity of the mobile spin label has increased by a factor of about 2. It should be pointed out here that even in the mobile spin-label case the three lines are not equally intense, the reason being that it does not have complete rotational freedom and the asymmetry of the lines is caused by the so-called anisotropic effects. Figure 6-15 shows the schematic drawing of the native and spin-labeled proteins. The interesting conclusion about the protein structure from this study is that the structure must be sufficiently open to admit the spin label for reaction with the $-SH$ group, and, on the other hand, the structure in the vicinity of the $-SH$ group must be sufficiently closed so as to immobilize the spin label. Thus, the protein structure has an elastic character.

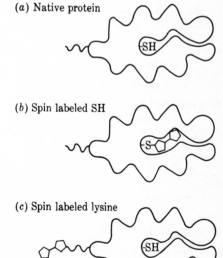

(a) Native protein

(b) Spin labeled SH

(c) Spin labeled lysine

Fig. 6-15 Schematic drawing of (a) native protein, (b) protein with labeled $-SH$ group yielding an immobile spin, and (c) protein with labeled lysine group yielding mobile spin. [*From O. H. Griffith and H. M. McConnell, Proc. Nat. Acad. Sci. U.S., **55**: 8 (1966). By permission of the National Academy of Sciences and the authors.*]

PROBLEMS

6-1. Calculate the Maxwell-Boltzmann distribution for electrons at 300°K and in a field of 3500 G.

6-2. Use McConnell's equation to calculate the unpaired spin density on the carbon atoms of the butadiene anion radical and hence construct the theoretical stick diagram (the coefficients for the atomic orbitals can be obtained from Chap. 2).

6-3. The hyperfine splitting constant for N^{14} in tetracyanoethylene anion radical is 1.56 G. Draw an energy level diagram to show the expected number of lines and intensities and hence construct a theoretical stick diagram.

6-4. Show that the Zeeman interaction for the proton is much less than that for the electron at 3500 G.

6-5. If ΔH is the linewidth at half-height of a Lorentzian line and ΔH_{ES} is the width between points of extreme slope of the first-derivative line, show that

$$\Delta H = \sqrt{3}\,\Delta H_{ES}$$

6-6. Consider the following reaction

$$N_2F_4 \rightleftharpoons 2\cdot NF_2$$

Suggest an ESR experiment to obtain the various thermodynamic quantities.

6-7. The statement that the ESR spectrum of the hydrogen atom consists of two lines of equal intensity is true only if we ignore the nuclear Zeeman interaction in Eq. (6-14). When this term is included, the Boltzmann distribution for the levels involved in the transition (see Fig. 6-6) will be slightly different. Calculate this small difference when the field is set at 3500 G.

SUGGESTIONS FOR FURTHER READING

Introductory

Carrington, A.: *Quart. Rev.*, **17**: 67 (1963).
Hecht, H. G.: "Magnetic Resonance Spectroscopy," John Wiley & Sons, Inc., New York, 1967
Ingram, D. J. E.: "Free Radicals as Studied by Electron Spin Resonance," Butterworth Scientific Publications, London, 1958.

Intermediate-advanced

Asycough, P. B.: "Electron Spin Resonance in Chemistry," Methuen and Co., Ltd., London, 1967.
Bersohn, M., and J. C. Baird: "An Introduction to Electron Paramagnetic Resonance," W. A. Benjamin, Inc., New York, 1966.
Carrington, A., and A. D. McLachlan: "Introduction to Magnetic Resonance," Harper & Row, Publishers, Incorporated, New York, 1967.
Poole, C. P.: "Electron Spin Resonance," Interscience Publishers, a division of John Wiley & Sons, Inc., New York, 1967.

READING ASSIGNMENTS

Structures of Organic Radicals, R. O. C. Norman, *Chem. Britain*, **6**: 66 (1970).
ESR Study of Organic Electron Transfer Reactions, R. Chang, *J. Chem. Educ.*, **47**: 563 (1970).

REFERENCES

1. See Bersohn and Baird, p. 205.
1a. For derivation of Eq. (6-16), see Bersohn and Baird, p. 214.
2. H. M. McConnell, *J. Chem. Phys.*, **28**: 1188 (1958).
3. G. N. Lewis and M. Kasha, *J. Am. Chem. Soc.*, **66**: 2100 (1944).
4. C. A. Hutchinson and B. W. Mangum, *J. Chem. Phys.*, **29**: 952 (1958).
5. E. Wasserman, L. C. Snyder, and W. A. Yager, *J. Chem. Phys.*, **41**: 1763 (1964).
6. T. Cole, D. E. Pritchard, N. Davidson, and H. M. McConnell, *Mol. Phys.*, **1**: 406 (1958).
7. W. T. Dixon and R. O. C. Norman, *J. Chem. Soc.*, 3119 (1963). For a general discussion of ESR studies of short-lived radicals, see R. O. C. Norman and B. C. Gilbert in V. Gold (ed.), "Advances in Physical Organic Chemistry," vol. 5, Academic Press, Inc., New York, 1967.
8. R. L. Ward and S. I. Weissman, *J. Am. Chem. Soc.*, **79**: 2086 (1957). For a comprehensive discussion of chemical rate processes and magnetic resonance, see C. S. Johnson, Jr., in J. S. Waugh (ed.), "Advances in Magnetic Resonance," vol. 1, p. 33, Academic Press, Inc., New York, 1965.
9. O. H. Griffith and H. M. McConnell, *Proc. Nat. Acad. Sci. U.S.*, **55**: 8 (1966). For a general discussion of applications of spin-labeling techniques to biomolecules, see O. H. Griffith and A. S. Waggoner, *Acc. Chem. Res.*, **2**: 17 (1969).

7

Microwave Spectroscopy

7-1 INTRODUCTION

Microwave spectroscopy deals with the pure rotational motion of molecules. The condition for observing resonance in this region is that a molecule must possess a *permanent* dipole moment. A rotating dipole generates an electric field which can interact with the electric component of the microwave radiation. If we assume that a diatomic molecule behaves like a rigid rotator, the rotational energy E_{rot} can be written as

$$E_{\text{rot}} = \frac{\hbar^2}{2I} J(J+1) \quad \text{ergs}$$

or

$$E_{\text{rot}} = \frac{\hbar}{4\pi cI} J(J+1)$$
$$= BJ(J+1) \quad \text{cm}^{-1} \tag{7-1}$$

where I is the moment of inertia, J the rotational quantum number, c

the velocity of light, and B, the rotational constant, is given by

$$B = \frac{\hbar}{4\pi cI} \quad cm^{-1} \tag{7-2}$$

THE SELECTION RULE

The quantum mechanical selection rule for a transition between any two rotational levels is given by

$$\mu_{nm} = \int \psi_n^* \mu \psi_m \, d\tau \tag{7-3}$$

where μ is the permanent dipole moment of the molecule. The dipole moment can be expressed by its three components along the x, y, and z axes.

$$\mu^2 = \mu_x{}^2 + \mu_y{}^2 + \mu_z{}^2 \tag{7-4}$$

In polar coordinates, these three components are given by

$$\begin{aligned}
\mu_x &= \mu \sin \theta \cos \phi \\
\mu_y &= \mu \sin \theta \sin \phi \\
\mu_z &= \mu \cos \theta
\end{aligned} \tag{7-5}$$

The transition dipole moment can now be written in three separate parts:

$$\begin{aligned}
\mu_x{}^{J'M'J''M''} &= \int_{\theta=0}^{\pi} \int_{\phi=0}^{2\pi} \psi^{J'M'} \mu \sin \theta \cos \phi \, \psi^{J''M''} \sin \theta \, d\theta \, d\phi \\
\mu_y{}^{J'M'J''M''} &= \int_{\theta=0}^{\pi} \int_{\phi=0}^{2\pi} \psi^{J'M'} \mu \sin \theta \sin \phi \, \psi^{J''M''} \sin \theta \, d\theta \, d\phi \\
\mu_z{}^{J'M'J''M''} &= \int_{\theta=0}^{\pi} \int_{\phi=0}^{2\pi} \psi^{J'M'} \mu \cos \theta \, \psi^{J''M''} \sin \theta \, d\theta \, d\phi
\end{aligned} \tag{7-6}$$

Where $\sin \theta \, d\theta \, d\phi = d\tau$, M is the component of J along a particular axis in space (for example, an external electric or magnetic field), the single prime indicates the upper and the double prime the lower state. In the absence of an external electric or magnetic field the selection rules are $\Delta J = \pm 1$ and $\Delta M = 0, \pm 1$

When a molecule is irradiated with the appropriate frequencies, the transition between any two successive rotational levels can occur. The energy difference ΔE_{rot} is given by

$$\Delta E_{rot} = BJ'(J' + 1) - BJ''(J'' + 1) \quad cm^{-1}$$

Since $J' = J'' + 1$, we have

$$\Delta E_{rot} = 2B J' \quad cm^{-1}$$

With a range of microwave frequencies present, the resulting spectrum will consist of a series of equally spaced lines with separation of $2B$ (Fig. 7-1).

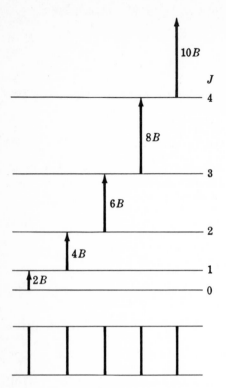

Fig. 7-1 Theoretical rotational spectrum for a rigid rotator.

INTENSITY

For a large assembly of molecules at thermal equilibrium, the ratio of number of molecules in an upper level J', n', to that in a lower level J'', n'', is given by Boltzmann's expression. However, this expression is modified in the present case because all the levels are *degenerate*, that is, for a given value of J, there are $2J + 1$ levels with the same energy. We write

$$\frac{n'}{n''} = \frac{g'}{g''}\frac{e^{-E'/kT}}{e^{-E''/kT}} = \frac{g'}{g''}\,e^{-\Delta E/kT} \tag{7-7}$$

where $g' = 2J' + 1$ and $g'' = 2J'' + 1$. The ratio of the number of molecules in a particular rotational level i, n_i, to the total number of molecules N, is given by

$$\frac{n_i}{N} = \frac{g_i e^{-E_i/kT}}{\displaystyle\sum_i g_i e^{-E_i/kT}} = \frac{g_i e^{-E_i/kT}}{Z_{\text{rot}}} \tag{7-8}$$

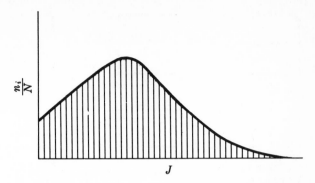

Fig. 7-2 Plot of the fractional number of molecules in a particular rotational level J versus J for a diatomic molecule at room temperature.

where Z_{rot} is the rotational partition function.[1]   Here it is clear that

$$\sum_i \frac{n_i}{N} = 1$$

Figure 7-2 shows the plot of the fractional number of molecules in a particular rotational level J versus J for a diatomic molecule at room temperature.   We see that some of the higher rotational levels are *more* populated than the lower ones.   There are two reasons for this behavior: (1) We always have $g' > g''$, and (2) the separation between successive rotational levels is small compared to the thermal energy kT at room temperature.   As a result the change in energy in going from one level to the next in some cases is not large enough to overweigh the change in the statistical weight.

7-2 EXPERIMENTAL TECHNIQUES

A microwave spectrometer consists of the following essential features:

1. Microwave radiation source, that is, klystron
2. Power supply
3. Frequency scanner to vary the frequency over the desired range
4. Sample cell
5. Detector and recorder

Figure 7-3 shows the block diagram of a typical microwave spectrometer. The overall setup and operation is somewhat similar to an ESR spectrometer—the important difference is that the frequency of the micro-

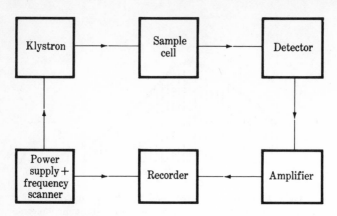

Fig. 7-3 Block diagram of a microwave spectrometer.

wave is varied either by electronic control or by a mechanical adjustment in this case. Also, no external magnetic field is required in a normal microwave experiment.

Most of the measurements are carried out on gaseous systems. The reason is that in liquids the typical number of molecular collisions is between 10^{12} and 10^{13} sec^{-1}, which is large compared to the reciprocal time for a complete rotation, about 10^{-10} sec. Consequently the rotational energy of the molecule will not be quantized and we can no longer apply Eq. (7-1). Similarly, the technique is inapplicable to solids, since in solids molecules are usually not free to rotate. However, in the gas phase at low pressures ($<10^{-2}$ mmHg) a molecule can execute many rotations before it collides with another molecule. Thus it is possible to obtain pure rotational spectra in this case. In order to increase the sensitivity at low pressures the sample tube is usually quite long—of the order of meters. This follows directly from Beer's law [Eq. (1-81)] that the absorbance is directly proportional to the path length of the cell. We shall be concerned only with the gaseous molecules at low pressures in this chapter.

7-3 THEORY

Equation (7-1) is correct only when applied to a rigid rotator. Since chemical bonds are nonrigid the interatomic distance will change somewhat as the molecule rotates because of the centrifugal force. Furthermore this "stretching" of the bond will depend on the frequency of rotation, or in quantum mechanics we say that it will depend on the value of J.† As a result the rotational lines are not exactly equally spaced, and the spacing between successive lines decreases with increasing J. As a first-order correction, the energy of a nonrigid rotator is given by

$$E_{\text{rot}} = BJ(J + 1) - D[J(J + 1)]^2 \qquad \text{cm}^{-1} \tag{7-9}$$

where D, the centrifugal distortion constant, is a small positive quantity of the order of one-thousandth part of B.

In addition to the centrifugal distortion a further correction has to be made because of the molecular vibration which changes the moment of inertia of the molecule. The correction for the rotational constant B_v in the vth vibrational state is

$$B_v = B_e - \alpha_e(v + \tfrac{1}{2}) \tag{7-10}$$

Where B_e is the rotational constant for the theoretical equilibrium state of no vibration, α_e is a constant about one-hundredth part of B_e. The energy E_{rot} now becomes

$$E_{\text{rot}} = B_eJ(J + 1) - D[J(J + 1)]^2 - \alpha_e(v + \tfrac{1}{2}) J(J,1) \quad \text{cm}^{-1} \tag{7-11}$$

CLASSIFICATION OF MOLECULES

The moment of inertia I of a rigid system about an axis through its center of mass is given by

$$I = \sum_i m_i r_i^2 \tag{7-12}$$

where r_i is the perpendicular distance from the mass m_i to the axis. For any given molecule, if we draw lines from its center of mass in a particular direction, of length proportional to the moment of inertia of the molecule about the line as an axis, the envelope of these lines will form an *ellipsoid*. Choosing the principal axes of the ellipsoid as x, y, and z, we write

$$\frac{x^2}{I_x^2} + \frac{y^2}{I_y^2} + \frac{z^2}{I_z^2} = 1 \tag{7-13}$$

where I_x, I_y, and I_z are the moments of inertia about these three axes (Fig. 7-4). We can now proceed to classify molecules in terms of the three moments of inertia according to Table 7-1.

† The rotational frequency has no physical meaning in quantum mechanics. It is only through the classical equation $P = I\omega$ (where P is the angular momentum and ω the angular frequency) that J is related to ν_{rot} ($\nu_{\text{rot}} = 2Bc \sqrt{J(J + 1)} \simeq 2BcJ$) in the limit of large values of J.

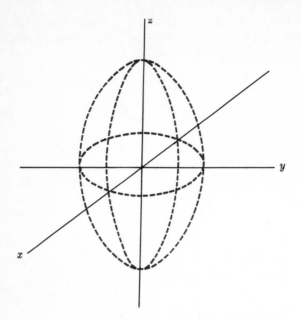

Fig. 7-4 An ellipsoid representing moments of inertia about a center of mass.

Figure 7-5 shows the rotation of different types of molecules about the three principal axes.

LINEAR MOLECULES

The rotational energy of any diatomic or linear polyatomic molecules is given by Eq. (7-1) or (7-9). Figure 7-5 shows that there is only one type

Table 7-1   Classification of molecules according to their moments of inertia

Moment of inertia†	Type of rotator	Example
$I_a = I_b$ $I_c = 0$	Linear	HF, OCS
$I_a = I_b = I_c$	Spherical top	Cubane, SF_6, UF_6, CH_4
$I_a = I_b > I_c$ $I_a = I_b < I_c$	Prolate $\Big\}$ symmetric top Oblate	NH_3, CH_3F
$I_a = I_b = I_c$	Asymmetric top	H_2O

† The principal axes of inertia are conventionally given the subscripts of a, b, and c, which may be thought of as the x, y, and z directions (see Fig. 7-5).

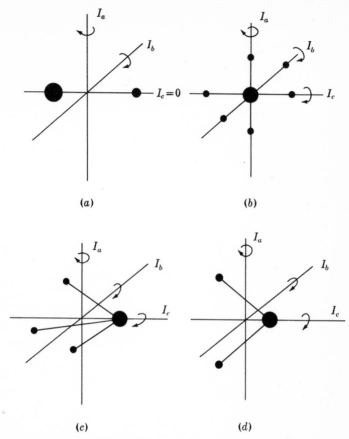

Fig. 7-5 Rotation of different types of molecules about the three principal axes. (*a*) Linear; (*b*) spherical top; (*c*) symmetric top; (*d*) asymmetric top.

of rotation possible. The rotation about the molecular axis will involve a much lower moment of inertia since only the electrons will contribute to the motion. Consequently, the very high frequency of the motion would correspond to an electronic transition. For diatomic molecules the separation between rotational lines gives the moment of inertia and hence the interatomic distance. For polyatomic molecules the moment of inertia is given in terms of more than one interatomic distance and the isotopic substitution technique is usually employed. We shall discuss this in more detail in Sec. 7-4.

SPHERICAL TOP MOLECULES

A spherical top molecule such as CH_4 or SF_6 normally does not possess a permanent dipole moment and therefore is microwave inactive.

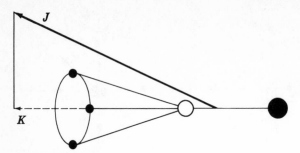

Fig. 7-6 Relation between **J** and **K** for a symmetric top molecule.

SYMMETRIC TOP MOLECULES

Classically the rotational energy of a symmetric top molecule is given by

$$E = \tfrac{1}{2}I_a\omega_a{}^2 + \tfrac{1}{2}I_b\omega_b{}^2 + \tfrac{1}{2}I_c\omega_c{}^2$$
$$= \frac{P_a{}^2}{2I_a} + \frac{P_b{}^2}{2I_b} + \frac{P_c{}^2}{2I_c} \tag{7-14}$$

where ω is the angular velocity and P is the angular momentum about the three axes.   According to quantum mechanics we have

$$P = \sqrt{J(J+1)}\,\hbar \tag{7-15}$$

and the corresponding energy[2]

$$E_{\text{rot}} = BJ(J+1) + K^2(A-B) \tag{7-16}$$

where

$$A = \frac{\hbar}{4\pi I_a c} \qquad \text{cm}^{-1}$$

$$B = \frac{\hbar}{4\pi I_b c} \qquad \text{cm}^{-1}$$

and K is the component of J about the unique axis (Fig. 7-6).   For every value of J there are $2J+1$ values of K given by

$$K = 0, \pm 1, \pm 2, \ldots, \pm J$$

The selection rules are

$$\Delta J = 0, \pm 1 \qquad \Delta K = 0$$

The correction for the centrifugal distortion is similar to that given for the linear molecule case.

ASYMMETRIC TOP MOLECULES

The analysis of these molecules is quite complex and there is no simple expression for the energy term. We shall not discuss this type of molecules here although the subject is treated in standard texts on microwave spectroscopy listed at the end of this chapter.

COUPLING OF NUCLEAR SPIN TO MOLECULAR ROTATION

It was pointed out in Chap. 4 that there are two ways to obtain the nuclear quadrupole coupling constant eQq. We can either perform a pure NQR experiment, or we can obtain the quantity from the pure rotational spectra. If a molecule contains a nucleus having a quadrupole moment, the interaction of the quadrupole with the electric field gradient due to the electrons couples together the nuclear spin I and the rotational angular momentum. The resultant angular moment is $\sqrt{F(F+1)}\,\hbar$ where F is given by

$$F = J + I, J + I - 1, \ldots, 0, \ldots, |J - I|$$

Since the nuclear quadrupole can have different orientations with respect to the electric field with different energies, each of the rotational levels will be split into $2I + 1$ levels if $J > I$ or $2J + 1$ levels if $I > J$. The energy for the system is given by[3]

$$E = eQq\left[\frac{\frac{3}{4}C(C+1) - I(I+1)J(J+1)}{2I(2I+1)2J(2J+1)}\right] \tag{7-17}$$

where

$$C = F(F+1) - I(I+1) - J(J+1)$$

It is important to note that Eq. (7-17) applies only to linear molecules containing one quadrupole nucleus. The difference between Eqs. (7-17) and (4-7) is that in solids molecules normally cannot rotate, and hence there is no coupling present. In the gas phase only the resultant angular momentum F is quantized with respect to the electric field whereas in solids the nuclear spin angular momentum is quantized. Figure 7-7 shows the splitting of the rotational levels as a result of the coupling. The selection rules for the transition are

$$\Delta J = \pm 1 \qquad \Delta F = 0, \pm 1$$

Thus it is possible to obtain the nuclear quadrupole coupling constant eQq from the fine structure of the rotational spectra. The microwave spectrum of HCN for the $J = 0 \to 1$ transition shows the expected splittings.[4] We note that both the *sign* and the magnitude of eQq can be obtained from the spectrum.

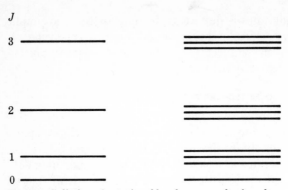

Fig. 7-7 Splitting of rotational levels as a result of nuclear quadrupole coupling. The energy levels on the left are without quadrupole field, and on the right, with quadrupole field. $I = 1$.

THE STARK EFFECT

When a rotational spectrum is recorded in the presence of a strong electric field ϵ the lines will in general be split and shifted. This is the Stark effect which was first observed in atomic spectra. Consider the following situation shown in Fig. 7-8. M is the projection of J along the external

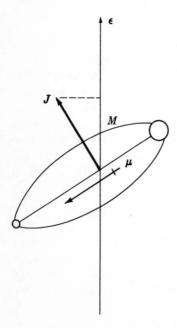

Fig. 7-8 A diatomic molecule in the presence of an external electric field.

electric field. The proper treatment of the Stark effect is beyond the scope of this book. Here we shall merely quote some of the important results. The energy in the presence of the field $E_{J,M}$ is given by[5]

$$E_{J,M} = E_J + a_{J,M}\epsilon + \tfrac{1}{2}b_{J,M}^2\epsilon^2 \tag{7-18}$$

where E_J is the energy of the rotator in the absence of the field and $a_{J,M}$ and $b_{J,M}$ are the first- and second-order Stark effects. If the dipole μ has a component along the direction of J we have the first-order Stark effect; if μ is perpendicular to J we have the second-order Stark effect, which is shown in Fig. 7-8. The difference between the first- and second-order Stark effect lies in the dependence of the energy on the electric field. For a linear molecule where there can be no component of μ in the direction of J, we have

$$E_{J,M} = \frac{\hbar}{4\pi cI} J(J + 1) + \tfrac{1}{2}b_{J,M}^2\epsilon^2 \tag{7-19}$$

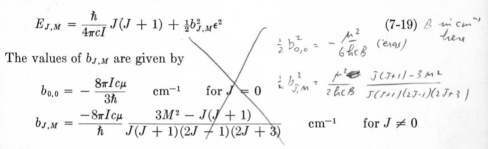

The values of $b_{J,M}$ are given by

$$b_{0,0} = -\frac{8\pi Ic\mu}{3\hbar} \quad \text{cm}^{-1} \quad \text{for } J = 0$$

$$b_{J,M} = -\frac{8\pi Ic\mu}{\hbar}\frac{3M^2 - J(J + 1)}{J(J + 1)(2J + 1)(2J + 3)} \quad \text{cm}^{-1} \quad \text{for } J \neq 0$$

Figure 7-9 shows the splitting of the rotational levels in the presence of ϵ.

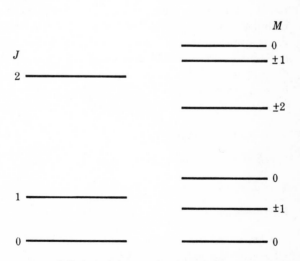

Fig. 7-9 Splitting of the rotational levels by an external electric field. The energy levels on the left are without external electric field, and on the right, with electric field.

The selection rules are†

$$\Delta J = \pm 1 \qquad \Delta M = 0$$

It can be shown from Eq. (7-19) that the shift in the $J = 0 \to 1$ line when the field is applied is given by $16\pi I c\mu/15\hbar$ and the $J = 1 \to 2$ line will be split into two with a separation of $32\pi I c\mu/105\hbar$. In each case the electric dipole moment of the molecule can be obtained. The Stark splittings of the $J = 1 \to 2$ transition for the linear OCS molecule has been studied and analyzed.[6]

7-4 APPLICATIONS

Structural determination   Information regarding molecular symmetry and molecular parameters can be obtained from microwave studies. As an example of the former let us consider the xenon oxytetrafluoride molecule $XeOF_4$.[7] The fact that the spectrum is characteristic of a symmetric top is consistent with the C_{4v} symmetry of the molecule.

$$\begin{matrix} F \\ F \\ F \\ F \end{matrix} \Bigg\rangle Xe = O$$

As Eq. (7-1) shows, the spacing between the rotational lines gives the moment of inertia I for a rigid rotator and hence the interatomic distance. For very accurate work the centrifugal distortion D must also be included. However, for low values of J the rigid rotator model usually holds quite well. Consider the linear molecule OCS. There are two different interatomic distances, and we are confronted with the difficulty of having to solve two unknowns in one equation. This difficulty can be dealt with by the isotopic technique as follows: Consider first the isotope species $O^{16}C^{12}S^{32}$ (Fig. 7-10). The center of mass is at O, so that

$$m_C r_1 + m_O r_2 = m_S r_3 \qquad (7\text{-}20)$$

The moment of inertia I is given by

$$I = m_C r_1{}^2 + m_O r_2{}^2 + m_S r_3{}^2 \qquad (7\text{-}21)$$

Furthermore we have

$$r_2 = r_{CO} + r_1 \qquad (7\text{-}22)$$
$$r_3 = r_{CS} - r_1 \qquad (7\text{-}23)$$

† We assume that ϵ is parallel to the electric field of the microwave,.as is usually the case.

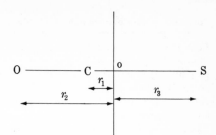

Fig. 7-10   The $O^{16}C^{12}S^{32}$ molecule.

where r_{CO} and r_{CS} are the interatomic distances.   Substituting Eqs. (7-22) and (7-23) into (7-20), we get

$$(m_C + m_O + m_S)r_1 = m_S r_{CS} - m_O r_{CO}$$

Hence

$$Mr_1 = m_S r_{CS} - m_O r_{CO}$$

or

$$r_1 = \frac{m_S r_{CS} - m_O r_{CO}}{M} \qquad (7\text{-}24)$$

where

$$M = m_C + m_O + m_S$$

Substituting Eqs. (7-22) and (7-23) into (7-21) we get

$$I = m_O(r_{CO} + r_1)^2 + m_C r_1^2 + m_S(r_{CS} - r_1)^2$$

and using the expression of r_1 in Eq. (7-24) we have

$$I = \frac{m_S r_{CS}^2 - m_O r_{CO}^2 + 2m_O m_S r_{CO} r_{CS} - m_S r_{CS}^2}{M} \qquad (7\text{-}25)$$

in which the unknowns are r_{CO} and r_{CS}.   If we assume that isotopic substitution does not alter the interatomic distance, we can obtain a similar expression for the moment of inertia I' of the molecule $O^{16}C^{12}S^{34}$.

$$I' = \frac{m'_S r_{CS}^2 - m_O r_{CO}^2 + 2m_O m'_S r_{CO} r_{CS} - m'_S r_{CS}^2}{M} \qquad (7\text{-}26)$$

In this way both I and I' can be obtained from the microwave spectra of $O^{16}C^{12}S^{32}$ and $O^{16}C^{12}S^{34}$ to give r_{CO} and r_{CS}.   Table 7-2 gives the interatomic distances obtained from the various isotopic substitutions.

Table 7-2   Interatomic distances in OCS

Pair of isotopic molecules	O—C (Å)	C—S (Å)
$O^{16}C^{12}S^{32}$, $O^{16}C^{12}S^{34}$	1.1647	1.5576
$O^{16}C^{12}S^{32}$, $O^{16}C^{13}S^{32}$	1.1629	1.5591
$O^{16}C^{12}S^{34}$, $O^{16}C^{13}S^{34}$	1.1625	1.5594
$O^{16}C^{12}S^{32}$, $O^{18}C^{12}S^{32}$	1.1552	1.5653

SOURCE: By permission of C. H. Townes and
A. L. Schawlow, "Microwave Spectroscopy,"
McGraw-Hill Book Company, New York, 1955.

It is important to note that the differences in interatomic distances
caused by changing isotopes that are shown in Table 7-2 are due entirely
to *zero-point* vibrations.   The vibrational motion of the molecule persists
at all temperatures including absolute zero.   The changes in the rotational
energy levels measured are usually associated with the ground vibrational
level, that is, the zero-point vibration.   However, due to this vibration
the bond lengths are slightly altered compared to the bond lengths in the
theoretical vibrationless state.   Furthermore, the amplitude of this zero-
point vibration is dependent on the mass of the atoms, and hence the
measured bond lengths will change upon isotopic substitution.   This
discrepancy can be corrected if rotational transitions can be observed for
different vibrational levels.   The result can then be extrapolated to the
"vibrationless" state.

Many of the nonlinear molecules have also been studied.   For
example, the structure of benzonitrile (C_6H_5CN) is found to have a
distorted hexagon.[8]

The inversion spectrum of ammonia   The microwave spectrum of
ammonia is of considerable interest.   Historically it was the first molecule
to be studied and also the first to show hyperfine structure due to nuclear
quadrupole coupling.[9a,b]   In addition each of the lines is split into a
doublet due to the inversion of the molecule.   Consider the inversion of
the nitrogen atom through the plane of the hydrogen atoms shown in
Fig. 7-11.   The new configuration cannot be obtained by the rotation of
the molecule; rather, it is the mirror image of the original configuration.
These two configurations are necessarily of the same energy, and a plot of
the potential energy curve[†] shows two minima with a hump in between.
The height of the hump, or more properly, the potential energy barrier,
represents the restriction to inversion.   For ammonia this potential
barrier is about 6 kcal mole^{-1}.   Classically it is impossible for the molecule

[†] We shall discuss the potential energy curve of molecules in more detail in Chap. 8.

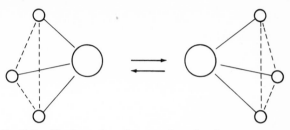

Fig. 7-11 The inversion of ammonia molecule.

to invert if it is in one of the first two vibrational levels. However, it is a well-known result of quantum mechanics[10] that particles can "tunnel" through barriers even though their energies are not great enough to get over them.

In principle this inversion is a vibrational motion. However, because the inversion is hindered, the interaction involves energy difference in the microwave region. The splitting of the vibrational levels shown in Fig. 7-12 is due to the *resonance interaction* between the two identical systems. This *inversion doubling* causes the rotational lines to split since the transitions are now between the two components of each of the vibrational levels.

It is interesting to point out here that analogous molecules such as PH_3 and AsH_3 do not show such doublings. Presumably the inversion frequencies in these systems are much slower due to the increase in the barrier height.

Internal rotation A closely related phenomenon to the inversion problem is the internal rotation in molecules. If a part of a molecule can rotate about a single bond then the internal potential energy of the molecule will

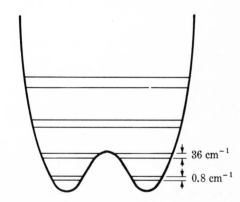

Fig. 7-12 Plot of the potential energy curve of ammonia molecule.

36 cm^{-1}

0.8 cm^{-1}

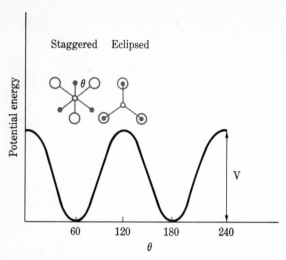

Fig. 7-13   Plot of the potential energy as a function of θ.

in general depend on the orientation of this part with the rest of the molecule.   Let us consider the 1,1,1-trifluoroethane molecule.   Figure 7-13 shows the plot of the potential energy vs. the different orientations of the CF_3 group.   The potential energy $V(\theta)$ is given by the simple expression

$$V(\theta) = \frac{V_3}{2}(1 - \cos 3\theta)$$

For low-potential barriers free rotation will take place.   On the other hand, for high barriers we will have "torsional vibrations" with quantum mechanical tunneling between equivalent conformations.   Table 7-3 lists some of the barrier heights measured by microwave spectroscopy.

Table 7-3   Barrier to rotation about single bond

Molecule	*Barrier (kcal mole^{-1})*
$CH_3OH(C\!-\!O)$	1.07
$CH_3OCH_3(C\!-\!O)$	2.72
$CH_3COOH(C\!-\!C)$	0.48
$CH_3NO_2(C\!-\!N)$	0.006
$CH_3SiH_3(C\!-\!Si)$	1.70
$C_6H_5CH_3(C\!-\!C)$	13.94
$N_2H_4(N\!-\!N)$	3.15

Finally we note that although a large number of barrier heights have been measured, no simple physical picture of the origin of such barriers has emerged.[11a,b]

Chemical analysis   An obvious example here is the determination of isotope abundance since each molecule has a unique moment of inertia depending on the particular nuclei present. The abundance can be obtained from the relative intensity of the lines or from the integrated area. The location of the position of the isotopic substitution within a molecule can also be determined by the microwave technique. For example the $J = 0 \rightarrow 1$ transitions for $N^{14}N^{15}O^{16}$ and $N^{15}N^{14}O^{16}$ molecules are found to be about 850 MHz apart.

In contrast to NMR, IR, and UV, microwave spectroscopy is not used routinely to identify compounds.

PROBLEMS

7-1. The rotational constant B is 10.6 cm^{-1} for H^1Cl^{35}.

   (a) Calculate the frequency in cm^{-1} for the pure rotational transition $J = 0 \rightarrow 1$.

   (b) Calculate the separation in cm^{-1} between the above transition and the corresponding one in H^1Cl^{37}.

   (c) Compare the result in (a) with the value kT at room temperature.

7-2. Use Eq. (7-8) to derive the value of E_i for the most populous state.

7-3. Obtain the transition frequency in terms of B and J for a symmetric top molecule.

7-4. The microwave spectrum of NH_3 shows a doublet of 0.8 cm^{-1}. How many times does the molecule invert every second?

7-5. The spacing between the rotational lines of HCl, assuming the rigid rotator model, is 20 cm^{-1}. Plot the theoretical rotational spectrum for the $J = 0 \rightarrow 1$ to $J = 14 \rightarrow 15$ transitions at room temperature.

SUGGESTIONS FOR FURTHER READING

Introductory

Barrow, G. M.: "The Structure of Molecules," W. A. Benjamin, Inc., New York, 1963.

Dunford, H. B.: "Elements of Diatomic Molecular Spectra," Addison-Wesley Publishing Company, Inc., Reading, Mass., 1968.

Intermediate-advanced

Barrow, G. M.: "Introduction to Molecular Spectroscopy," McGraw-Hill Book Company, New York, 1962.

Sugden, T. M., and Kenney, C. N.: "Microwave Spectroscopy of Gases," D. Van Nostrand Company, Inc., Princeton, N.J., 1965.

Townes, C. H., and Schawlow, A. L.: "Microwave Spectroscopy," McGraw-Hill Book Company, New York, 1955.

READING ASSIGNMENTS

Microwave Spectroscopy, W. H. Kirchhoff, *Chem. Eng. News*, **46**: 82 (1968).
Microwave Spectroscopy in the Undergraduate Laboratory, R. H. Schwendeman, H. N. Voltrauer, V. W. Laurie, and E. C. Thomas, *J. Chem. Educ.*, **47**: 526 (1970).

REFERENCES

1. D. F. Eggers, Jr., N. W. Gregory, G. D. Halsey, Jr., and B. S. Rabinovitch, "Physical Chemistry," p. 160, John Wiley & Sons, Inc., New York, 1964.
2. Townes and Schawlow, p. 50.
3. Townes and Schawlow, p. 151.
4. J. W. Simons, W. E. Anderson, and W. Gordy, *Phys. Rev.*, **77**: 77 (1950), **86**: 1055 (1952).
5. Townes and Schawlow, p. 248.
6. T. W. Dakin, W. E. Good, and D. K. Coles, *Phys. Rev.*, **70**: 560 (1946).
7. J. Martins and E. B. Wilson, Jr., *J. Chem. Phys.*, **41**: 570 (1964).
8. B. Bak, D. Christensen, W. B. Dixon, L. Hansen-Nygaard, and J. Rastrup-Andersen, *J. Chem. Phys.*, **37**: 2027 (1962).
9a. C. E. Cleeton and N. H. Williams, *Phys. Rev.*, **45**: 234 (1934).
9b. W. E. Good, *Phys. Rev.*, **70**: 213 (1946).
10. F. L. Pilar, "Elementary Quantum Chemistry," p. 92, McGraw-Hill Book Company, New York, 1968.
11a. E. B. Wilson, Jr., The Problem of Barriers to Internal Rotation in Molecules, in I. Prigogine (ed.), *Advan. Chem. Phys.*, **2**: 367 (1959).
11b. J. P. Lowe, Barriers to Internal Rotation about Single Bonds, in A. Streitwieser, Jr., and R. W. Taft (eds.), *Prog. Phys. Org. Chem.*, **6**: 1 (1968).

8
Infrared and Raman Spectroscopy

ORIGIN OF INFRARED TRANSITIONS

Infrared (IR) spectroscopy is usually divided into three regions:

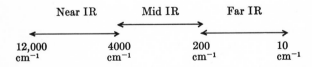

The far-IR region deals with the pure rotational motion of the molecules. Because of the experimental difficulties in the generation and detection at low frequencies far IR is seldom used in chemical spectroscopy. In this chapter we shall discuss only the mid- and near-IR regions.

If we assume that a diatomic molecule behaves like a simple harmonic oscillator, then the energy of vibration E_{vib} is given by

$$E_{\text{vib}} = (v + \tfrac{1}{2})h\nu \tag{8-1}$$

The transition electric-dipole-moment integral μ_{nm} is given by

$$\mu_{nm} = \int \psi_n^* \mu \psi_m \, d\tau \tag{8-2}$$

where ψ_n and ψ_m are the upper and lower vibrational wave functions an μ is the electric dipole moment. For pure rotational motion μ is a con stant. For vibrational motion μ will change as a result of the change c bond length if the molecule has a permanent dipole moment to start witl Thus the dipole moment of a vibrating molecule can be represented a

$$\mu(r) = \mu_0 + \left(\frac{d\mu}{dr}\right)_0 q + \frac{1}{2}\left(\frac{d^2\mu}{dr^2}\right)_0 q^2 + \cdots \tag{8-3}$$

where μ_0 is the permanent electric dipole moment, and $q = r - r_e$ wher r_e is the equilibrium bond distance. Figure 8-1 shows a typical varia tion of μ with r. Substituting Eq. (8-3) into (8-1) and neglecting all bu the first two terms in Eq. (8-3) we obtain†

$$\mu_{nm} = \int \psi_n^* \left[\mu_0 + \left(\frac{d\mu}{dr}\right)_0 q \right] \psi_m \, d\tau$$

$$= \int \psi_n^* \left(\frac{d\mu}{dr}\right)_0 q \psi_m \, d\tau \tag{8-4}$$

It is clear that there will be no transition if the molecule does not chang its dipole moment during a vibration. This immediately excludes a homonuclear diatomic molecules for IR studies. The selection rule for simple harmonic vibrator is just $\Delta v = \pm 1$ (Sec. 1-3).

† ψ_m and ψ_n are orthonormal functions.

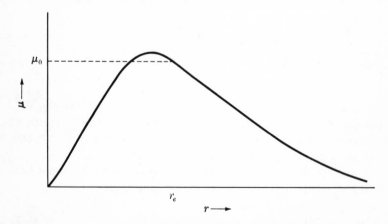

Fig. 8-1 A typical variation of dipole moment with internuclear distance.

The ratio of the population in level v, N_v, to the total population N s given by

$$\frac{N_v}{N} = \frac{e^{-E_v/kT}}{\sum\limits_{i}^{N} e^{-E_i/kT}} = \frac{e^{-E_v/kT}}{Z_{\text{vib}}} \tag{8-5}$$

where Z_{vib} is the vibrational partition function.[1] For most diatomic molecules at room temperature the population of the excited levels is small compared with that of the ground level, and since IR experiments are absorptions, we would expect only a single line. However, as we shall see in Sec. 8-3, this is hardly the case.

3-2 EXPERIMENTAL TECHNIQUES OF IR SPECTROSCOPY

Figure 8-2 shows the schematic diagram of an IR spectrometer. The source is usually a Nernst glower (electrically heated rare earth oxides). Since the incident radiation is not monochromatic, a monochromator is required to spread out the continuous emission so that only narrow regions of the spectrum pass through the slit to the detector. The dispersing

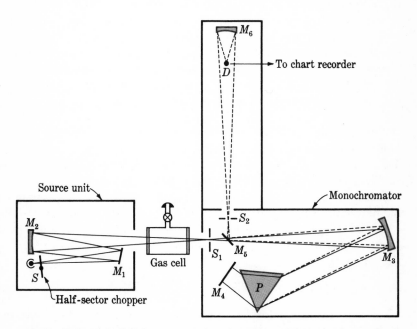

Fig. 8-2 Schematic diagram of an IR spectrometer. (*By permission of D. P. Shoemaker and C. W. Garland, "Experiments in Physical Chemistry," McGraw-Hill Book Company, New York, 1962.*)

element is usually a prism or a grating. A thermocouple is used as the detector.

Compounds are usually recorded in the gas or liquid phase. The resolution is worse for the latter because of the molecular collisions which broaden out the rotational fine structures (to be discussed later).

8-3 APPEARANCE OF IR SPECTRA

DIATOMIC MOLECULES

If the higher terms in Eq. (8-3) cannot be neglected then we must consider integrals of the type

$$\left(\frac{d^2\mu}{dr^2}\right)_0 \int \psi_n^* q^2 \psi_m \, d\tau \qquad \text{etc.}$$

The inclusion of these higher terms can lead to the selection rules $\Delta v = \pm 1$, $\pm 2, \ldots$. However, since

$$\left(\frac{d\mu}{dr}\right)_0 \gg \left(\frac{d^2\mu}{dr^2}\right)_0 \gg \left(\frac{d^3\mu}{dr^3}\right)_0 \cdots$$

the additional lines will in general be very weak. Thus we have the strong line corresponding to the $v = 0 \to 1$ transition, that is, the *fundamental* transition and the additional weak lines corresponding to $v = 0 \to 2$, $0 \to 3, \ldots$ transitions which are called *overtones*. If the spectrum is recorded above room temperature, we can have the so-called *hot bands* due to the $v = 1 \to 2$, $2 \to 3, \ldots$ transitions since at higher temperatures the higher levels will be somewhat populated.

Another way in which the selection rules $v = 0 \to 2$, $0 \to 3, \ldots$ can arise is from the anharmonicities in the potential function. Real molecules cannot behave like simple harmonic oscillators since molecules dissociate at large values of r and acquire very large amounts of energy as r decreases. Figure 8-3 shows the potential energy curves for a simple harmonic oscillator and a diatomic molecule. The equation for the simple harmonic oscillator is given by $V(r) = \frac{1}{2}k(r - r_e)^2$ [see Eq. (1-53) in which x is replaced by $(r - r_e)$]. Of all the formulas proposed to fit the curve for diatomic molecules the one by Morse is best known.[2] It has the following form:

$$V(r) = D_e[1 - e^{-a(r-r_e)}]^2 \tag{8-6}$$

where a is a constant for a given molecule in a specific electronic state, D_e is the spectroscopic dissociation energy, and D_0 (see Fig. 8-3) is the chemical dissociation energy. The difference between D_e and D_0 gives the zero-point energy. From Eq. (8-6) it is clear that as $r \to \infty$, $V(\infty) = D_e$; at $r = r_e$, $V(r_e) = 0$.

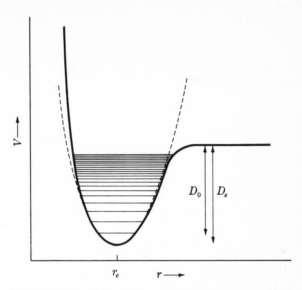

Fig. 8-3 Potential energy curve for a diatomic molecule. The dotted line is for a simple harmonic oscillator.

Since the spacing between the rotational energy levels is much smaller than that between the vibrational levels, it is possible to obtain *pure* rotational spectra as discussed in Chap. 7.   However, it is not possible to obtain pure vibrational spectra since transitions between any two vibrational levels will normally always be accompanied by the simultaneous change in rotational energy.   Thus it is quite easy to detect the rotational fine structure in well-resolved vibrational spectra.   The important consequence is that we can no longer treat the rotation and vibration as separate, independent motions [see Eq. (1-85)].

SELECTION RULES

The transition dipole-moment integral for a *vibrating rotator* is given by

$$\mu^{v'v''} = \int \psi_{vr}^{v'*} \mu \psi_{vr}^{v''}\, d\tau \qquad (8\text{-}7)$$

where ψ_{vr} is given by[3]

$$\psi_{vr} = R(r)\Theta(\theta)\Phi(\phi)$$
$$= \psi_{\text{vib}} \frac{r - r_e}{r} \psi_{\text{rot}}(\theta,\phi)$$

Equation (8-7) is expanded to give

$$\mu^{v'J'M',\ v''J''M''} = \int \psi_{\text{vib}}^{v'*} \mu(r) \psi_{\text{vib}}^{v''}\, dr \int_\theta \int_\phi \psi_{\text{rot}}^{J'M'*}\, \mu(\theta,\phi) \psi_{\text{rot}}^{J''M''}$$
$$\sin\theta\, d\theta\, d\phi \quad (8\text{-}8)$$

In polar coordinates we have

$$\mu(\theta,\phi,r) = \mu(r)\mu(\theta,\phi) \tag{8-9}$$

Neglecting the anharmonicity of μ or the vibration we obtain the following selection rules:

$$\Delta v = \pm 1 \qquad \Delta J = \pm 1 \qquad \Delta M = 0, \pm 1$$

In considering the energy of the molecule executing rotational and vibrational motions, we should keep the following points in mind:

1. Due to centrifugal distortion the energy of a rotator should be written as

$$E_{\text{rot}} = BJ(J + 1) - DJ^2(J + 1)^2$$

2. The rotational constant B as given in Eq. (7-2) is proportional to $1/r^2$. However, the average value of this quantity, $<1/r^2>$, is not equal to $1/r_e^2$, where r_e is the equilibrium distance (see Fig. 8-3). Thus the value of B and also that of D will depend on the amplitude of vibration, that is, the vibrational quantum number v. We write

$$B_v = B_e - \alpha(v + \tfrac{1}{2}) \tag{8-10}$$
$$D_v = D_e + \beta(v + \tfrac{1}{2}) \tag{8-11}$$

where α and β are small constants, the subscript v denotes the vth vibrational state, and e the hypothetical vibrationless state. The D_e in Eq. (8-11) should not be confused with the D_e shown in Fig. 8-3.

3. Since molecules are not simple harmonic oscillators, corrections must be made for the anharmonicity. We have

$$E_{\text{vib}} = (v + \tfrac{1}{2})h\nu_e - x(v + \tfrac{1}{2})^2 h\nu_e \qquad \text{ergs}$$

or

$$E_{\text{vib}} = (v + \tfrac{1}{2})\bar\nu_e - x(v + \tfrac{1}{2})^2\bar\nu_e \qquad \text{cm}^{-1} \tag{8-12}$$

where x is a small positive constant and $\bar\nu_e$ is the fundamental frequency $[\bar\nu_e = (1/2\pi)\sqrt{k/\mu}]$. The energy for the vibration-rotation interaction E_{vr} is then given by

$$E_{vr} = B_vJ(J + 1) - D_vJ^2(J + 1)^2$$
$$+ (v + \tfrac{1}{2})\bar\nu_e - x(v + \tfrac{1}{2})^2\bar\nu_e \tag{8-13}$$

Each line in a vibration-rotation spectrum generally corresponds to the change of the molecule from one particular rotational (and vibrational) level to another. Neglecting anharmonicities and centrifugal distortions we have

$$\Delta E_{vr} = B_{v'}J'(J' + 1) + (v' + \tfrac{1}{2})\bar\nu_e - B_{v''}J''(J'' + 1) - (v'' + \tfrac{1}{2})\bar\nu_e$$

Since $v' = v'' + 1$ and $J' = J'' + 1$ we have

$$\Delta E_{vr} = \bar{\nu}_e + 2B_{v'} + (3B_{v'} - B_{v''})J + (B_{v'} - B_{v''})J^2$$

If we assume that $B_{v'} = B_{v''} = B$ then

$$\Delta E_{vr} = \bar{\nu}_e + 2B + 2BJ \tag{8-14}$$

where J is now the rotational quantum number of the lower level.

A vibration-rotation spectrum is divided into the so-called P and R branches as follows:

P branch $(J' = J'' - 1)$: $\Delta E_{vr} = \bar{\nu}_e - 2BJ$
R branch $(J' = J'' + 1)$: $\Delta E_{vr} = \bar{\nu}_e + 2B(J + 1)$

Figure 8-4 shows the vibration-rotation energy level diagram for a diatomic molecule. Figure 8-5 shows the IR spectrum of HBr.

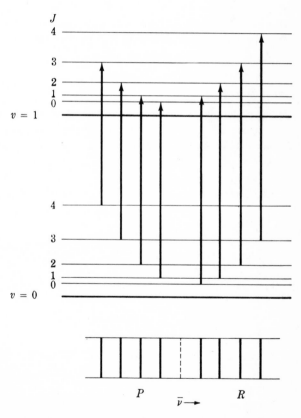

Fig. 8-4 Vibrational-rotational energy level diagram for a diatomic molecule.

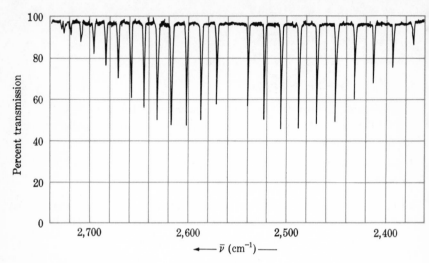

Fig. 8-5 IR spectrum of HBr. (*By permission of G. M. Barrow, "Introduction to Molecular Spectroscopy," McGraw-Hill Book Company, New York, 1962.*)

LINEAR MOLECULES

Linear molecules, whether they possess permanent electric dipole moment or not, are IR active if some of their vibrations produce an oscillating dipole. This is most clearly seen in the case of carbon dioxide. There are two types of IR active vibrations:

1. The oscillating dipole moment is parallel to the molecular axis

$$\vec{O}\cdots\overset{\leftarrow}{C}\cdots\cdots\vec{O} \qquad\qquad \overset{\leftarrow}{O}\cdots\cdots\vec{C}\cdots\overset{\leftarrow}{O}$$

   This gives rise to the so-called parallel band. The selection rules are $\Delta v = \pm 1$ and $\Delta J = \pm 1$.
2. The oscillating dipole moment is perpendicular to the molecular axis

$$\overset{\uparrow}{O}\cdots\cdots\overset{\uparrow}{C}\cdots\cdots O \qquad\qquad \overset{\oplus}{O}\cdots\cdots\overset{\ominus}{C}\cdots\cdots\overset{\oplus}{O}$$
$$\downarrow \qquad\qquad \downarrow$$

   This gives rise to the perpendicular band. The selection rules are $\Delta v = \pm 1$ and $\Delta J = 0, \pm 1$. The $\Delta J = 0$ transition arises because a component of the electric field ϵ perpendicular to the molecular axis can excite the vibration without causing rotational changes. For a diatomic molecule this component of ϵ cannot cause vibration unless the molecule possesses a nonzero electronic orbital momentum, for example, nitric oxide. The $\Delta J = 0$ transition gives rise to a new group of lines called the Q branch.

Q branch $(J' = J'')$: $\quad \Delta E_{vr} = \bar{\nu}_e(v' - v'') = \bar{\nu}_e$

SYMMETRIC TOP MOLECULES

The energy due to the vibration-rotation interaction, neglecting anharmonicity and centrifugal distortion, is given by

$$\Delta E_{vr} = (v + \tfrac{1}{2})\bar{\nu}_e + B_v J(J + 1) + (A - B)K^2 \qquad \text{cm}^{-1}$$

where A, B, and K are defined in Sec. 7-3. The selection rules are

Parallel band: $\qquad \Delta K = 0 \qquad \Delta J = 0, \pm 1$
Perpendicular band: $\quad \Delta K = \pm 1 \qquad \Delta J = 0, \pm 1$

Figure 8-6 shows the parallel band of CH_3Br.

ASYMMETRIC MOLECULES

The treatment of asymmetric molecules is quite complex and will not be dealt with here. The reader is referred to the standard texts listed at the end of the chapter.

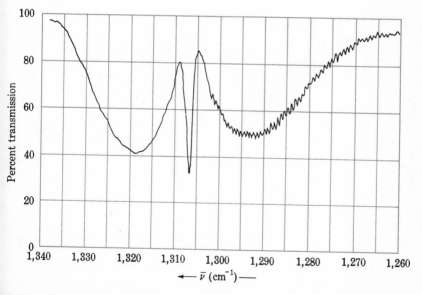

Fig. 8-6 The parallel band of CH_3Br. *(By permission of G. M. Barrow, "Introduction to Molecular Spectroscopy," McGraw-Hill Book Company, New York, 1962.)*

SPIN STATISTICS AND SYMMETRY

The vibration-rotation spectra of many symmetric linear molecules show alternation in intensities (Fig. 8-7). This is due to the symmetry of the wave function of the molecule as follows: First let us consider the total wave function of a homonuclear diatomic molecule. We have

$$\psi_{total} = \psi_e \psi_{vib} \psi_{rot} \psi_n$$

where ψ_e and ψ_n are the electronic and nuclear wave functions. According to the Pauli exclusion principle (Chap. 2) ψ_e must be antisymmetric to the exchange of the coordinates of a pair of electrons. Actually this principle applies to all the fermions. On the other hand, the wave function of bosons is symmetric to the exchange of the coordinates. For a given homonuclear diatomic molecule ψ_{total} must be either symmetric or antisymmetric with respect to the exchange of the two nuclei depending on whether they are bosons or fermions. Both ψ_e and ψ_{vib} are symmetric to the exchange of the two nuclei since the former does not depend on the nuclear coordinates and the latter depends only on the internuclear distance. However, ψ_{rot} will be either symmetric and antisymmetric to this exchange depending on whether the rotational quantum number J is even or odd. This arises from the fact that the exchange of the nuclei in the cartesian coordinates can be presented by (x,y,z)

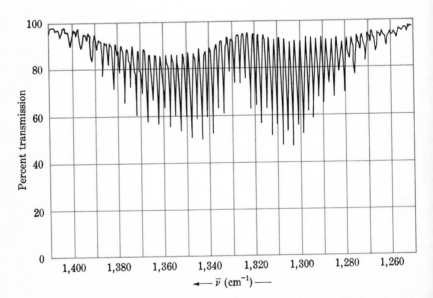

Fig. 8-7 IR spectrum of C_2H_2. (*By permission of G. M. Barrow, "Introduction to Molecular Spectroscopy," McGraw-Hill Book Company, New York, 1962.*)

$\rightarrow (-x, -y, -z)$, or, in the polar coordinates, $(r, \theta, \phi) \rightarrow (r, \pi - \theta, \pi + \phi)$. It can be readily shown that the substitution of $\pi - \theta$ for θ and $\pi + \phi$ for ϕ in the rotational wave functions (Sec. 1-2) gives the following results

$$\psi_{\text{rot}} \xrightarrow{\text{exchange}} -\psi_{\text{rot}} \qquad \text{if } J = 1, 3, 5, \ldots$$

$$\psi_{\text{rot}} \xrightarrow{\text{exchange}} \psi_{\text{rot}} \qquad \text{if } J = 0, 2, 4, \ldots$$

If we consider a specific molecule, say hydrogen, we can construct the nuclear spin wave functions as follows: $\alpha(1)\alpha(2)$, $\alpha(1)\beta(2)$, $\beta(1)\alpha(2)$, $\beta(1)\beta(2)$ where (1) and (2) denote the two nuclei. $\alpha(1)\alpha(2)$ and $\beta(1)\beta(2)$ are obviously symmetric to the exchange but $\alpha(1)\beta(2)$ and $\beta(1)\alpha(2)$ are neither symmetric nor antisymmetric to the exchange. Thus we have to take linear combinations of them as follows

$$\left.\begin{array}{l} \alpha(1)\alpha(2) \\ \alpha(1)\beta(2) + \alpha(2)\beta(1) \\ \beta(1)\beta(2) \end{array}\right\} \quad \begin{array}{l} \text{symmetric nuclear} \\ \text{wave functions } \psi_n{}^s \end{array}$$

$$\alpha(1)\beta(2) - \alpha(2)\beta(1) \qquad \begin{array}{l} \text{antisymmetric} \\ \text{nuclear wave function } \psi_n{}^a \end{array}$$

Since protons are fermions, ψ_{total} must be antisymmetric to the exchange of the nuclei. Thus ψ_{total} can be written either as

$$\psi_{\text{total}} = \psi_e \psi_{\text{vib}} \psi_n{}^a \psi_{\text{rot}} \qquad J = 0, 2, 4, \ldots$$

or

$$\psi_{\text{total}} = \psi_e \psi_{\text{vib}} \psi_n{}^s \psi_{\text{rot}} \qquad J = 1, 3, 5, \ldots$$

Consequently levels with the rotational wave functions having odd J values will have a statistical weight three times that with even J values because of the difference in $\psi_n{}^s$ and $\psi_n{}^a$.[†]

Although the hydrogen molecule itself is inactive in IR, the differences in statistical weights for rotational levels do show up in the vibration-rotation spectra of symmetric linear polyatomic molecules such as CO_2 and C_2H_2. Figure 8-8 shows the rotational energy level diagram for C_2H_2. In the vibrational ground state, ψ_{vib} is an even function. ψ_{rot} with odd values of J will have three times the statistical weight of ψ_{rot} with even J's. In the first excited vibrational state ψ_{vib} is an odd function so that the statistical weight for the rotational levels is reversed.

† These two possible combinations of wave functions give rise to the ortho- and para-hydrogen molecules.

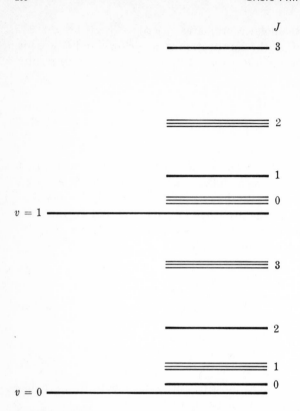

Fig. 8-8 Rotational energy level diagram for C_2H_2.

8-4 ORIGIN OF RAMAN SPECTROSCOPY

In addition to the absorption experiments just discussed for IR a great deal of information can also be obtained from the scattered radiation. We shall now discuss this phenomenon below.

QUANTUM THEORY OF LIGHT SCATTERING

When a beam of light shines on an assembly of molecules the photons of energy $h\nu$ (assuming monochromatic radiation) collide with the molecules and one of the two things can happen.† If the collision is elastic the deflected photons, that is, the scattered radiation, will have the same energy as the incident photons. On the other hand if the collision is inelastic, the deflected photons will have either higher or lower energy than the incident photons. As a good approximation we can say that the

† We assume no photochemical reactions would occur.

kinetic energy of the photon and the molecule remains the same before and after the collision. The law of conservation of energy then requires that

$$h\nu + E = h\nu' + E' \qquad (8\text{-}15)$$

where E represents the rotational, vibrational, and electronic energy of the molecule before collision and E' represents the same values after collision. Rearranging Eq. (8-15) we get

$$\frac{E - E'}{h} = \nu - \nu'$$

The scattered radiation is classified as

$$E = E' \; (\nu = \nu') \qquad \text{Rayleigh scattering}$$

$$\left.\begin{array}{l} E > E' \; (\nu < \nu') \\ E < E' \; (\nu > \nu') \end{array}\right\} \qquad \text{Raman scattering}$$

Thus, in the Raman-scattering case, energy is either given to or taken away from the molecule as a result of the interaction. Figure 8-9 shows the energy level diagrams for these interactions.

When a molecule is promoted from the ground state to a higher, unstable vibrational state by the incident radiation it can either return to the original state or to a different vibrational state. The former gives rise to Rayleigh scattering and the latter to Raman scattering which, in this case, gives rise to the *Stokes* lines. If a molecule is initially in the first excited vibrational state it can be promoted to a higher, unstable state and then return to the ground state. This is also Raman scattering which gives rise to the *anti-Stokes* lines. The intensity of the Rayleigh line is much stronger than that of the Stokes lines which in turn are much stronger than the anti-Stokes lines. Of course we also have changes in the rotational levels in addition to the vibrational level transitions; hence the appearance of the fine structures in Fig. 8-9b. The choice of frequency of the incident radiation for a Raman experiment is not so restrictive as that for IR as long as the frequency is not high enough to induce electronic transitions.

CLASSICAL THEORY OF LIGHT SCATTERING

Although not strictly correct, the classical theory does give a good physical picture of the Raman effect. When a molecule is subject to an oscillating electric field ϵ of the type

$$\epsilon = \epsilon_0 \sin 2\pi\nu t \qquad (8\text{-}16)$$

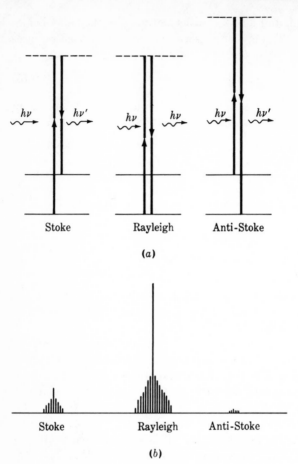

Fig. 8-9 (a) The interactions that give rise to the Stokes, anti-Stokes, and Rayleigh lines. (b) The theoretical spectra.

the *induced* oscillating dipole moment μ is proportional to the strength of the electric field. The constant of this proportionality is called the polarizability (α). That is,

$$\mu \propto \epsilon$$

or

$$\mu = \alpha\epsilon \tag{8-17}$$

Note that α is not merely a constant of proportionality. If a molecule is not electronically isotropic the magnitude of the induced dipole will be different along the electric field component. Thus α is a *tensor* quantity.

We write

$$\mu_x = \alpha_{xx}\epsilon_x + \alpha_{xy}\epsilon_y + \alpha_{xz}\epsilon_z$$
$$\mu_y = \alpha_{yx}\epsilon_x + \alpha_{yy}\epsilon_y + \alpha_{yz}\epsilon_z \qquad (8\text{-}18)$$
$$\mu_z = \alpha_{zx}\epsilon_x + \alpha_{zy}\epsilon_y + \alpha_{zz}\epsilon_z$$

where

$$\epsilon^2 = \epsilon_x{}^2 + \epsilon_y{}^2 + \epsilon_z{}^2 \qquad (8\text{-}19)$$

The physical significance of the α_{ij} elements in Eq. (8-18) is that the i subscript denotes the direction of the induced dipole moment by the oscillating electric field component in the j direction. The polarizability tensor is symmetric, that is, $\alpha_{ij} = \alpha_{ji}$. It is possible in this case to transform the system into the principal axes system so that all α_{ij}'s are zero if $i \neq j$, and we have

$$\mu_{x'} = \alpha_{x'x'}\epsilon_{x'}$$
$$\mu_{y'} = \alpha_{y'y'}\epsilon_{y'} \qquad (8\text{-}20)$$
$$\mu_{z'} = \alpha_{z'z'}\epsilon_{z'}$$

where the primes indicate the new coordinate system. If we draw lines in any direction from the origin of length proportional to $1/\sqrt{\alpha}$, then the locus of the points of the lines will form a surface known as the polarizability ellipsoid whose principal axes are x', y', and z'. The equation of the ellipsoid is given by

$$\alpha_{x'x'}x'^2 + \alpha_{y'y'}y'^2 + \alpha_{z'z'}z'^2 = 1 \qquad (8\text{-}21)$$

If $\alpha_{x'x'} = \alpha_{y'y'} = \alpha_{z'z'}$, the molecule is said to be isotropic.

Since the rotational and vibrational motion continuously change the electronic distribution of the molecule, the polarizability can be written as a series:

$$\alpha = \alpha_e + \left(\frac{\partial \alpha}{\partial Q}\right)_0 Q + \frac{1}{2}\left(\frac{\partial^2 \alpha}{\partial Q^2}\right)_0 Q^2 + \cdots \qquad (8\text{-}22)$$

where α_e is the polarizability of the molecule in its equilibrium position, and $Q = r - r_e$, where r is the internuclear distance at a given instant. Figure 8-10 shows a typical variation of α with r.

If the molecule is vibrating with frequency ν_0, then Q itself must be a function of time. We write†

$$Q = Q_0 \sin 2\pi\nu_0 t \qquad (8\text{-}23)$$

Neglecting all but the first two terms in Eq. (8-22) we have

$$\alpha = \alpha_e + \left(\frac{\partial \alpha}{\partial Q}\right)_0 Q_0 \sin 2\pi\nu_0 t \qquad (8\text{-}24)$$

† We assume simple harmonic motion.

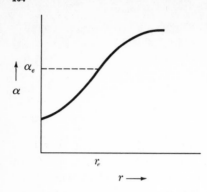

The induced dipole moment is†

$$\mu = \epsilon_0 \alpha_e \sin 2\pi\nu t + \epsilon_0 \left(\frac{\partial \alpha}{\partial Q}\right)_0 Q_0 \sin 2\pi\nu_0 t \sin 2\pi\nu t$$

$$= \epsilon_0 \alpha_e \sin 2\pi\nu t + \frac{1}{2}\epsilon_0 Q_0 \left(\frac{\partial \alpha}{\partial Q}\right)_0$$

$$[\cos 2\pi(\nu - \nu_0)t - \cos 2\pi(\nu + \nu_0)t] \quad (8\text{-}25)$$

Thus the classical theory correctly predicts that the oscillating dipole has frequencies at ν (Rayleigh scattering) and $\nu \pm \nu_0$ (Raman scattering). The intensity of the Rayleigh line is proportional to α_e, and all the Raman lines are proportional to $(\partial\alpha/\partial Q)_0$. This is incorrect since the Stokes lines are much more intense than the anti-Stokes lines.

8-5 EXPERIMENTAL TECHNIQUES OF RAMAN SPECTROSCOPY

Unlike IR, the detecting system for the Raman experiment is at right angles to the incident radiation. Figure 8-11 shows the schematic

† We have made use of the relationship $\sin A \sin B = \frac{1}{2}\cos (A - B) - \frac{1}{2}\cos (A + B)$ in deriving Eq. (8-25).

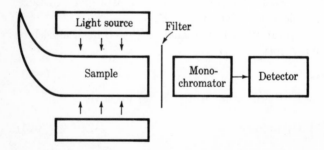

Fig. 8-11 Schematic diagram for a Raman spectrometer.

diagram of a Raman spectrometer. The exciting source is usually a mercury discharge tube whose blue line at 4358 Å is most useful for the scattering experiment. The detector is either a photographic plate or a photomultiplier. The horn shape of the cell reduces the direct reflection of the source from the back of the cell. Before 1966 Raman spectroscopy was not used for routine analytical work, as was the case with IR. The reason for this was that the scattered radiation was so weak in intensity that it was not unusual to take several hours or even days to record a spectrum. However, with the use of laser (see Chap. 14) as the exciting source, the scope of Raman spectroscopy is greatly widened. The advantages of using laser are:

1. A single, intense frequency source replaces the multiple-lined mercury lamp. No filtering is necessary.
2. The resolution is better since the linewidth of a laser line is smaller than that of a mercury-excited line.
3. Laser light is highly coherent; hence, the greater ease in focusing and collimating the radiation.
4. A large choice of exciting frequencies are available, and it is possible to study colored solutions without causing any electronic transitions. This is of great practical importance in studying inorganic salts in solution since many of them are colored compounds. The use of IR is ruled out since the solvent is usually water, which absorbs strongly in the IR region.

8-6 APPEARANCE OF RAMAN SPECTRA

PURE ROTATION SPECTRA

The rotational energy for a linear molecule is

$$E_{rot} = BJ(J + 1) - D[J(J + 1)]^2 \quad cm^{-1}$$

The centrifugal distortion constant D is usually neglected since the resolution of the Raman spectra is worse than that of the microwave spectra. The selection rules are $\Delta J = 0, \pm 2$, where the zero corresponds to the Rayleigh scattering and the ± 2 corresponds to the Raman transitions. The factor 2 arises since during a complete rotation the polarizability ellipsoid rotates *twice* as fast as the molecule. The $\Delta J = 2$ transitions give rise to the Stokes lines; the $\Delta J = -2$ transitions, the anti-Stokes lines. The Stokes and anti-Stokes lines in the pure rotation spectra are of comparable intensity. It should be noted that any linear molecule belonging either to the $C_{\infty h}$ or $D_{\infty h}$ point group has a rotational Raman spectrum.

The rotational energy change in a Raman transition is given by

$$\Delta E_{rot} = BJ'(J' + 1) - BJ''(J'' + 1)$$
$$= B(4J + 6)$$

where J denotes the lower rotational level. Figure 8-12 shows the theoretical spectrum of a diatomic molecule. Figure 8-13 shows the argon-laser-excited pure rotational spectrum of nitrogen.

The rotational energy of a symmetric top molecule is given by

$$E_{rot} = BJ(J + 1) + (A - B)K^2 \qquad cm^{-1}$$

The selection rules are

$$\begin{aligned}\Delta K &= 0 \\ \Delta J &= \pm 2\end{aligned} \qquad \text{if } K = 0$$

and

$$\begin{aligned}\Delta K &= 0 \\ \Delta J &= 0, \pm 1, \pm 2 \qquad \text{if } K \neq 0\end{aligned}$$

The treatment of asymmetric top molecules is quite complex and will not be dealt with here.

Finally we note that spherical top molecules such as CH_4 or UF_6 do not give rise to pure rotational Raman spectra. The polarizability ellipsoid of these molecules is a spherical surface, and the rotation of this ellipsoid will not produce any change in polarizability.

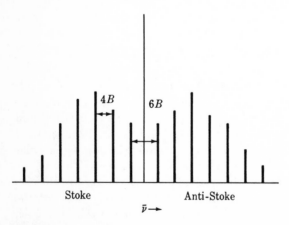

Fig. 8-12 Theoretical Raman rotational spectrum of a diatomic molecule.

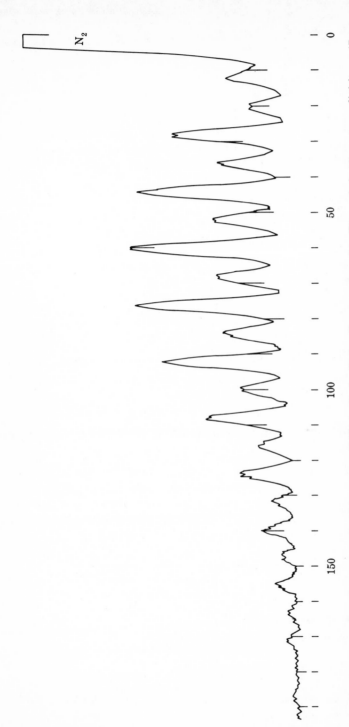

Figure 8-13 Laser-excited pure rotational spectrum of N_2 (only the Stokes lines are shown). The scale is 10 cm^{-1} per division. (*By permission of Jarrell–Ash, Division of Fisher Scientific Co.*)

167

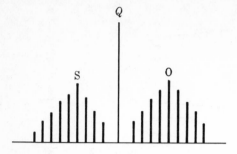

Fig. 8-14 Theoretical vibrational-rotational Raman spectrum of a linear molecule.

VIBRATION–ROTATION SPECTRA

The energy of a linear vibrating rotator is given by

$$E_{vr} = (v + \tfrac{1}{2})\bar{\nu}_e - x(v + \tfrac{1}{2})^2\bar{\nu}_e + BJ(J + 1)$$

The selection rules are $\Delta J = 0, \pm 2$ and $\Delta v = \pm 1$. Consider only the Stokes lines ($v = 0 \rightarrow 1$ transitions): We have

$\Delta J = 0$	$\Delta E_{vr} = \bar{\nu}_e(1 - 2x)$	Q branch
$\Delta J = 2$	$\Delta E_{vr} = \bar{\nu}_e(1 - 2x) + B(4J + 6)$	S branch
$\Delta J = -2$	$\Delta E_{vr} = \bar{\nu}_e(1 - 2x) - B(4J + 6)$	O branch

Figure 8-14 shows the theoretical vibration-rotation Raman spectrum of a linear molecule. The resolution of mercury-excited Raman spectra is usually poorer than that of IR spectra.

8-7 GROUP THEORY AND NORMAL MODES OF VIBRATION OF POLYATOMIC MOLECULES

Consider a molecule having N atoms: If we specify the position of each atom in space with the coordinates x, y, and z, then there will be $3N$ coordinates for the entire system (molecule). Since the molecule has translational, rotational, and vibrational motion, these $3N$ coordinates can be assigned to each of the motions in the following way:

Motion	Degrees of freedom† required to describe the motion
Translation (center of mass)	3
Rotation (about the center of mass)	2 (linear)
	3 (nonlinear)
Vibration	$3N - 5$ (linear)
	$3N - 6$ (nonlinear)

† The number of degrees of freedom possessed by a molecule is the number of coordinates required to completely specify the position of the nuclei.

The vibrational motion of a polyatomic molecule may seem very complex and irregular. However, careful analyses have shown that there are certain basic vibrations that can set the molecule into a periodic oscillation in which the nuclei all move in phase, that is, they all pass through the mean position at the same time. These unique vibrations are called *normal modes* or *normal modes of vibration*. They are of essential importance in vibrational analysis since any arbitrary vibration of the molecule can be described by a superposition of the normal modes. Depending on the particular molecule these normal modes are active either in IR or Raman or both.

NORMAL MODE ANALYSIS

Only a brief outline will be given here. The reader is referred to the text by Barrow for greater details.

The potential energy V of a molecule can be given in the form of a Taylor series

$$V = V_0 + \sum_i \left(\frac{\partial V}{\partial q_i}\right)_0 q_i + \frac{1}{2} \sum_i \sum_j \left(\frac{\partial^2 V}{\partial q_i \, \partial q_j}\right)_0 q_i q_j + \cdots \qquad (8\text{-}26)$$

where $q_i, q_j, \ldots$ are the coordinates of nuclei $i, j, \ldots$, and the subscript zero indicates the equilibrium state of the molecule. Since V_0 is the potential in the equilibrium state and therefore a constant, we can arbitrarily call it zero. The term $(\partial V/\partial q_i)_0$ corresponds to the minimum of the potential energy curve and is therefore also set to be zero. Neglecting all the higher terms except one in Eq. (8-26), we have

$$\begin{aligned} V &= \frac{1}{2} \sum_i \sum_j \left(\frac{\partial^2 V}{\partial q_i \, \partial q_j}\right)_0 q_i q_j \\ &= \frac{1}{2} \sum_i \sum_j b_{ij} q_i q_j \end{aligned} \qquad (8\text{-}27)$$

where

$$b_{ij} = \left(\frac{\partial^2 V}{\partial q_i \, \partial q_j}\right)_0$$

Similarly, the kinetic energy T of the molecule can also be written as a series. Thus

$$\begin{aligned} T &= \tfrac{1}{2}(a_{11}\dot{q}_1{}^2 + a_{22}\dot{q}_2{}^2 + \cdots + 2a_{12}\dot{q}_1\dot{q}_2 \\ &\qquad + \cdots + 2a_{13}\dot{q}_1\dot{q}_3 + \cdots + 2b_{23}\dot{q}_2\dot{q}_3 + \cdots) \\ &= \tfrac{1}{2} \sum_i \sum_j a_{ij}\dot{q}_i\dot{q}_j \end{aligned} \qquad (8\text{-}28)$$

where $\dot{q} = dq/dt$, and the a_{ij}'s and b_{ij}'s are functions of the mass of atoms.

Equations (8-27) and (8-28) can be used to construct the hamiltonian to be used in the Schrödinger equation. However, the cross products of the type $q_i q_j$ prevent the separation of variables. This problem can be solved by using the normal coordinates Q such that

$$
\begin{aligned}
q_1 &= c_{11}Q_1 + c_{12}Q_2 + c_{13}Q_3 + \cdots \\
q_2 &= c_{21}Q_1 + c_{22}Q_2 + c_{23}Q_3 + \cdots \\
&\cdots\cdots\cdots\cdots\cdots\cdots\cdots\cdots \\
q_i &= c_{i1}Q_1 + c_{i2}Q_2 + c_{i3}Q_3 + \cdots
\end{aligned}
\tag{8-29}
$$

By suitable choice of the coefficients c_{ik} of this linear transformation it is possible to write T and V in terms of the new coordinates as

$$
\begin{aligned}
T &= \tfrac{1}{2}(\dot{Q}_1{}^2 + \dot{Q}_2{}^2 + \cdots + \dot{Q}_{3n}{}^2) \\
&= \sum_i \dot{Q}_i{}^2
\end{aligned}
\tag{8-30}
$$

and

$$
\begin{aligned}
V &= \tfrac{1}{2}(\lambda_1 Q_1{}^2 + \lambda_2 Q_2{}^2 + \cdots + \lambda_{3N} Q_{3N}{}^2) \\
&= \tfrac{1}{2}\sum_i \lambda_i Q_i{}^2
\end{aligned}
\tag{8-31}
$$

where

$$
\lambda_i = 4\pi^2 \nu_i{}^2
$$

We can now proceed to solve the Schrödinger time-independent equation of the vibration problem.   For each normal coordinate Q_i, that is, for each normal mode, we have

$$
\mathcal{3C}_i = (T_i + V_i)_{\text{operator}}
$$

Thus we have

$$
-\frac{\hbar^2}{2}\frac{\partial^2 \psi(Q_i)}{\partial Q_i{}^2} + \frac{\lambda_i}{2} Q_i{}^2 \psi(Q_i) = E_i \psi(Q_i)
\tag{8-32}
$$

The sum of the total vibrational energy is given by

$$
E_{\text{vib}} = E_1 + E_2 + \cdots + E_{3N-5,6} = \sum^{3N-5,6} E_i
\tag{8-33}
$$

where $E_i = (v + \tfrac{1}{2})h\nu_i$.   The total wave function is given by

$$
\begin{aligned}
\psi_{\text{vib}} &= \psi_v(Q_1)\psi_v(Q_2) \cdots \psi_v(Q_{3N-5,6}) \\
&= \prod_i^{3N-5,6} \psi_v(Q_i)
\end{aligned}
\tag{8-34}
$$

where the subscript v denotes the particular vibrational level and i the particular normal mode.   From Sec. 1-2 we have

$$
\psi_v(\xi) = N_v H_v(\xi) e^{-\xi^2/2}
$$

Writing Q_i for ξ we have for the ground state

$$
\psi_0(Q_i) = \left(\frac{1}{\alpha_i \pi}\right)^{\frac{1}{4}} \exp \frac{-\tfrac{1}{2}Q_i{}^2}{\alpha_i}
\tag{8-35}
$$

and for the first excited state

$$\psi_1(Q_i) = \left(\frac{1}{\alpha_i{}^3\pi}\right)^{\frac{1}{4}} Q_i \exp \frac{-\frac{1}{2}Q_i{}^2}{\alpha_i} \tag{8-36}$$

Thus by assuming the simple harmonic oscillator model we have broken the vibrational motion of the molecule down into $3N - 5$ or $3N - 6$ simple harmonic oscillators in the normal coordinates Q_i.

If we consider only the $v = 0 \rightarrow 1$ transition then there will be $3N - 5$ fundamental frequencies for the linear and $3N - 6$ fundamental frequencies for the nonlinear molecules. The fundamental frequency for the ith normal mode is given by

$$\nu_i = \frac{1}{2\pi} \sqrt{\frac{k}{\mu}}$$

where k is the force constant and μ is now some complicated function of masses. Note that k is no longer associated with a single bond as in the diatomic case, and there is no straightforward correlation between k and the strength of the chemical bond. Table 8-1 shows the force constants for some typical bonds obtained spectroscopically.

Table 8-1 Frequencies of fundamental vibrational transitions ($v = 0 \rightarrow 1$) and the bond force constants calculated from these data

Molecule	$\bar{\nu}$ (cm^{-1})	k (dynes cm^{-1})
H_2	4159.2	5.2×10^5
D_2	2990.3	5.3
HF	3958.4	8.8
HCl	2885.6	4.8
HBr	2559.3	3.8
HI	2230.0	2.9
CO	2143.3	18.7
NO	1876.0	15.5
F_2	892	4.5
Cl_2	556.9	3.2
Br_2	321	2.4
I_2	213.4	1.7
O_2	1556.3	11.4
N_2	2330.7	22.6
Li_2	246.3	1.3
Na_2	157.8	1.7
NaCl	378	1.2
KCl	278	0.8

SOURCE: By permission of G. M. Barrow, "Introduction to Molecular Spectroscopy," McGraw-Hill Book Company, New York, 1962.

SELECTION RULES

We have already seen that the most intense transitions, that is, the fundamental transitions, are $v = 0 \rightarrow 1$. The problem now is to decide which normal modes will be active in IR and/or Raman. This means that we need to know whether μ or α will change in a vibration. We need to investigate integrals of the type

$$\int \psi_i{}^1 \mu \psi_i{}^0 \, d\tau \qquad \text{and} \qquad \int \psi_i{}^1 \alpha \psi_i{}^0 \, d\tau$$

(where the subscript i denotes the ith normal mode) in order to determine their behavior with respect to the symmetry operations of the molecule. The basic principle we apply here is that an operation which transforms a molecule into an indistinguishable configuration cannot change any of the observable properties of the molecule. If such an operation results in a change of the sign of one of the integrals, the integral must be identically equal to zero.

The symmetry properties of the integral can be readily determined since we are provided with the following information: (1) The symmetry properties of ψ_{vib} are known. (2) The components of μ transform like translations (x,y,z). (3) The components of α transform like products of translations (xx,yy,zz,xy,xz,yz). For the $v = 0 \rightarrow 1$ transition we see that $\psi_i{}^0$ is totally symmetric and $\psi_i{}^1$ transforms like Q_i [see Eq. (8-36)]. One obvious approach here is to solve the complete vibrational problem to obtain the Q_i's and then investigate their symmetry behavior. Since their symmetry behavior is identical with that of the wave functions of the complete hamiltonian, Q_i's must be the bases for the irreducible representations of the molecular point group. This immediately suggests the use of group theory to determine the symmetry behavior of the normal modes without any detailed numerical calculations.

The symmetry elements discussed in Chap. 2 are based on the assumption that molecules are rigid bodies. Since molecules are never completely at rest the symmetry elements of any molecule therefore apply to the situation when all the atoms are in their equilibrium position. The procedure here is to choose some convenient way to represent any distortion of the molecule, for example, displacements in the x, y, and z directions for each atom or changes in the bond length and angle. Since any distortion can be expressed by a linear combination of normal coordinates, our arbitrary representation must contain the irreducible representations which correspond to the normal coordinates.

Let us consider the SO_2 molecule which has the C_{2v} symmetry (Fig. 8-15). The total number of coordinates is $3N = 9$, which is also the number of degrees of freedom. To apply group theory we need only the characters of the set of matrices which represent the operations of C_{2v}. The dimensions of *all* the matrices are $3N \times 3N$. For the identity

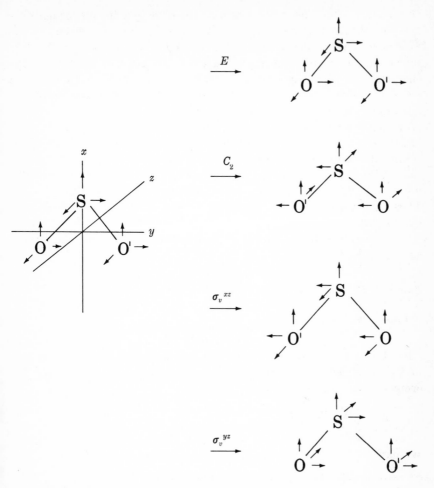

Fig. 8-15 Symmetry operations on the nine vectors of SO_2.

operation E we have

$$
\begin{bmatrix} x_S \\ y_S \\ z_S \\ x_O \\ y_O \\ z_O \\ x_{O'} \\ y_{O'} \\ z_{O'} \end{bmatrix} = E \begin{bmatrix} x_S \\ y_S \\ z_S \\ x_O \\ y_O \\ z_O \\ x_{O'} \\ y_{O'} \\ z_{O'} \end{bmatrix} = \begin{bmatrix} 1 & 0 & 0 & 0 & 0 & 0 & 0 & 0 & 0 \\ 0 & 1 & 0 & 0 & 0 & 0 & 0 & 0 & 0 \\ 0 & 0 & 1 & 0 & 0 & 0 & 0 & 0 & 0 \\ 0 & 0 & 0 & 1 & 0 & 0 & 0 & 0 & 0 \\ 0 & 0 & 0 & 0 & 1 & 0 & 0 & 0 & 0 \\ 0 & 0 & 0 & 0 & 0 & 1 & 0 & 0 & 0 \\ 0 & 0 & 0 & 0 & 0 & 0 & 1 & 0 & 0 \\ 0 & 0 & 0 & 0 & 0 & 0 & 0 & 1 & 0 \\ 0 & 0 & 0 & 0 & 0 & 0 & 0 & 0 & 1 \end{bmatrix} \begin{bmatrix} x_S \\ y_S \\ z_S \\ x_O \\ y_O \\ z_O \\ x_{O'} \\ y_{O'} \\ z_{O'} \end{bmatrix}
$$

Hence $\chi(E) = 9$. Actually we do not even have to write out the matrices for E, C_2, $\sigma_v{}^{xz}$, and $\sigma_v{}^{yz}$ since we can determine the diagonal elements alone. Let us consider in detail the symmetry operations on the nine vectors shown in Fig. 8-15. The following rules are useful in evaluating the characters: (1) If a displacement coordinate (x, y, or z) is left unchanged by a symmetry operation it contributes $+1$ to the character of the operation. (2) If a displacement coordinate changes sign by a symmetry operation it contributes -1 to the character of the operation. (3) If a displacement coordinate changes position by a symmetry operation it contributes zero to the character of the operation. For the C_2 operation we have $x_S = -x_S(-1)$, $y_S = -y_S(-1)$, $z_S = z_S(1)$, $x_O = -x_{O'}(0)$, $y_O = -y_{O'}(0)$, $z_O = z_{O'}(0)$, $x_{O'} = -x_O(0)$, $y_{O'} = y_O(0)$, and $z_{O'} = z_O(0)$. Thus we get $\chi(C_2) = -1$. Similarly we can show that $\chi(\sigma_v{}^{xz}) = 1$ and $\chi(\sigma_v{}^{yz}) = 3$.

The representation for the $3N$ coordinates, Γ_{3N}, can now be written as

Element	E	C_2	$\sigma_v{}^{xz}$	$\sigma_v{}^{yz}$
Γ_{3N}	9	-1	1	3

Since Γ_{3N} is a reducible representation, our next task is to find out how many times each of the irreducible representations (A_1, A_2, B_1, and B_2) occur in Γ_{3N}. To do this we use Eq. (2-9)

$$n_j = \frac{1}{h} \sum_R \chi_j(R)\chi(R)$$

We have

$$n_{A_1} = \tfrac{1}{4}[9 \times 1 + (-1) \times 1 + 1 \times 1 + 3 \times 1] = 3$$
$$n_{A_2} = \tfrac{1}{4}[9 \times 1 + (-1) \times 1 + 1 \times (-1) + 3 \times (-1)] = 1$$
$$n_{B_1} = \tfrac{1}{4}[9 \times 1 + (-1) \times (-1) + 1 \times 1 + 3 \times (-1)] = 2$$
$$n_{B_2} = \tfrac{1}{4}[9 \times 1 + (-1) \times (-1) + 1 \times (-1) + 3 \times 1] = 3$$

Hence

$$\Gamma_{3N} = 3A_1 + A_2 + 2B_1 + 3B_2$$

The next step is to sort out the different types of motion (translation, rotation, and vibration) connected with the irreducible representations. From the third column (from left) of the C_{2v} character table we can immediately write down

$$\Gamma_{\text{trans}} = A_1 + B_1 + B_2$$
$$\Gamma_{\text{rot}} = A_2 + B_1 + B_2$$

Therefore we have

$$\Gamma_{\text{vib}} = \Gamma_{3N-6} = 2A_1 + B_2$$

Hence there are three normal modes for SO_2 and they transform like A_1 and B_2.†

We are now ready to consider the selection rules. Consider first the IR case. For the $v = 0 \to 1$ transition it is required that at least one of the following three integrals be nonzero:

$$\int \psi_i{}^1 x \psi_i{}^0 \, d\tau \qquad \int \psi_i{}^1 y \psi_i{}^0 \, d\tau \qquad \int \psi_i{}^1 z \psi_i{}^0 \, d\tau$$

Since $\psi_i{}^0 \psi_i{}^1$ transforms like Q_i, a transition will be allowed if the product $\psi_i{}^0 \psi_i{}^1$ is the basis of a representation of the point group which contains x, y, and z. For SO_2 it is clear that all the normal modes are active in IR. For the Raman case we can similarly show that for an allowed transition the normal mode must belong to the same representation as one of the components of the polarizability tensor, x^2, y^2, z^2, xy, xz, or yz. Here again all the normal modes are active in Raman.

DEPOLARIZATION RATIO

Additional information regarding the symmetry of normal modes can often be obtained from Raman experiments if we can determine the degree of depolarization, ρ_n, of the scattered light.

If a molecule is excited by unpolarized light, the scattering at right angles to the incident light will be at least partially polarized. The degree of polarization depends on the way in which the polarizability ellipsoid varies during the vibration. It is also possible, and in fact more convenient, to excite molecules using polarized light, for example, polarized laser light. Figure 8-16 shows the incident and scattered radiation. The incident light is along the x axis, and the direction of observation is perpendicular to the x axis in the yz plane. ρ_n is defined as

$$\rho_n = \frac{I_\perp}{I_\parallel} \tag{8-37}$$

where $I_\perp$ is the intensity of the scattered light parallel to the xy plane and $I_\parallel$ is that parallel to the yz plane. The maximum value of ρ_n is $\frac{6}{7}$. A Raman line is characterized as polarized or depolarized according to the value of ρ. We have

$$\rho = 0 - \tfrac{6}{7} \qquad \text{polarized}$$
$$= \tfrac{6}{7} \qquad \text{depolarized}$$

† If a vibration is symmetric with respect to the highest-fold rotation axis it is designated by A, and if the vibration is antisymmetric about the same axis it is designated by B. The letter E, which is absent here, is for doubly degenerate vibrations.

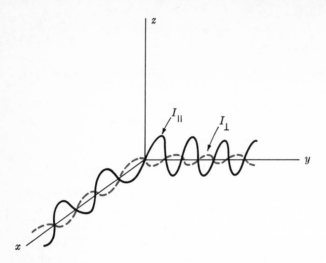

Fig. 8-16 The parallel and perpendicular components of the scattered radiation. The incident radiation is along the x axis and the molecule is situated at the origin.

Only totally symmetric vibrations can yield polarized Raman lines.[5] A reading of $\frac{6}{7}$ is of course ambiguous, but usually we can determine whether these lines are polarized or depolarized. Figure 8-17 shows the three normal modes of SO_2. Figure 8-18 shows the polarizability changes during the same normal modes of vibrations.

RULE OF MUTUAL EXCLUSION

Consider a linear triatomic molecule such as CO_2. Since the molecule is linear there are $3N - 5 = 4$ normal modes of vibration, which are shown in Fig. 8-19. Figure 8-20 shows the polarizability changes during the same normal modes of vibrations. The fundamental frequencies of SO_2 and CO_2 are listed in Table 8-2.

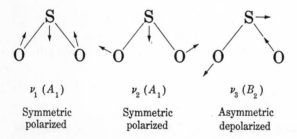

$\nu_1 (A_1)$ $\nu_2 (A_1)$ $\nu_3 (B_2)$

Symmetric   Symmetric   Asymmetric
polarized   polarized   depolarized

Fig. 8-17 The three vibrational modes of SO_2.

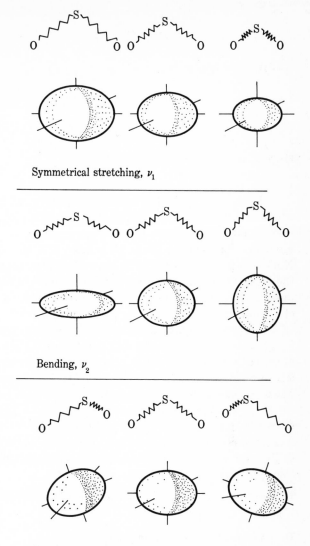

Symmetrical stretching, ν_1

Bending, ν_2

Antisymmetrical stretching, ν_3

Fig. 8-18 Polarizability changes of SO_2 during the normal modes of vibrations. [*From R. S. Tobias, J. Chem. Educ.,* **44:** *40 (1967). By permission of the Journal of Chemical Education.*]

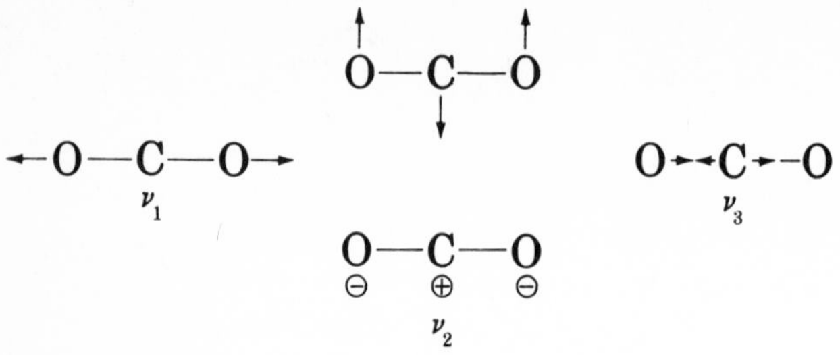

$$O-C-O$$

$$\leftarrow O-C-O \rightarrow \qquad\qquad O \rightarrow\!\!\leftarrow C \rightarrow\!\!-O$$
$$\nu_1 \qquad\qquad\qquad\qquad\qquad \nu_3$$

$$O-C-O$$
$$\ominus \quad \oplus \quad \ominus$$
$$\nu_2$$

Fig. 8-19   Normal modes of vibrations of CO_2.

In the SO_2 case all the vibrations are active in the IR and Raman whereas the same does not hold for CO_2. This can be accounted for by the mutual exclusion rule: If a molecule has a center of symmetry then any vibration that is active in the IR is inactive in the Raman and vice versa. Note that it is also possible to have vibrations that are active in neither the IR nor the Raman. In general whenever the same vibration appears in both IR and Raman it is fairly certain that the molecule lacks a center of symmetry.

The above result can be deduced from group theory as follows: The CO_2 molecule belongs to the $D_{\infty h}$ point group. The inversion of any one of the cartesian coordinates x, y, or z about the center of symmetry changes the sign of that coordinate. Hence all representations generated by x, y, or z must belong to a u representation. However, products such as xy, yz, xz, x^2, y^2, or z^2, which represent components of the polarizability tensor, do not change sign and belong to the g representations.† Thus from the $D_{\infty h}$ character table we see that vibrations belonging to the u representations are active in the IR and vibrations belonging to the g representations are active in the Raman.

† The symbols g and u stand for gerade (even) and ungerade (odd).

Table 8-2   Fundamental frequencies of vibration

Molecule	Frequency (cm^{-1})
SO_2	$\nu_1 = 1150$ (IR, Raman)
	$\nu_2 = 520$ (IR, Raman)
	$\nu_3 = 1360$ (IR, Raman)
CO_2	$\nu_1 = 1300$ (Raman)
	$\nu_2 = 667$ (IR)
	$\nu_3 = 2350$ (IR)

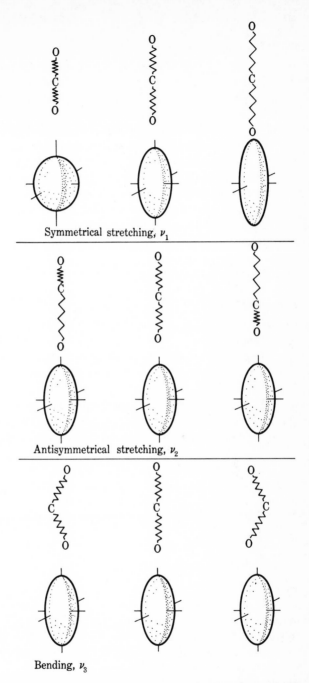

Symmetrical stretching, ν_1

Antisymmetrical stretching, ν_2

Bending, ν_3

Fig. 8-20   Polarizability changes of CO_2 during the normal modes of vibrations.   [*From R. S. Tobias, J. Chem. Educ.,* **44:** *40 (1967).   By permission of the Journal of Chemical Education.*]

OVERTONES AND COMBINATION BANDS

In addition to the lines due to the normal modes there are usually more lines observed in an IR or Raman spectrum. Since no molecular vibration is strictly harmonic, there will be transitions given by $\Delta v = \pm 2$, ± 3, . . . which are called overtones, as mentioned earlier. Overtone bands are generally much weaker than the fundamental bands and can be easily identified since they appear at or near 2ν, 3ν, . . . , where ν is the fundamental frequency.

Molecular vibrations are not only anharmonic but they are also not completely independent of one another. This means that it is possible to excite two or more modes *simultaneously* to give rise to the so-called combination bands at $2\nu_1 + \nu_2$, $\nu_1 + \nu_2 + \nu_3$, $\nu_3 - \nu_2$, etc.

It is important to note here that both overtones and combination bands have to satisfy symmetry requirements in a manner similar to that discussed for the normal modes of vibration.[6]

FERMI RESONANCE

One of the fundamental frequencies of $CO_2(\nu_1)$ is at 1300 cm^{-1}, and one of the combination bands ($2\nu_2$) is at 1334 cm^{-1}. However, the Raman spectrum shows two lines at 1285 and 1388 cm^{-1} instead of the two expected frequencies. This shift in resonance frequency is known as Fermi resonance.[7]

According to quantum mechanics whenever two states ψ_1 and ψ_2 (having energies E_1 and E_2) belong to the same irreducible representation, there will be a *resonance interaction*, the energy of which E_{inter} is given by

$$E_{\text{inter}} = \int \psi_1 \mathcal{3C} \psi_2 \, d\tau$$

where $\mathcal{3C}$ is the hamiltonian operator for the interaction. This interaction "mixes" the two original states to give ψ_1' and ψ_2' whose energies are

$$E_1' = E_1 + E_{\text{inter}}$$
$$E_2' = E_2 - E_{\text{inter}}$$

where $E_1 > E_2$.

The case just mentioned for CO_2 is an example of such interaction. We see that ν_1 is shifted down from 1300 to 1285 cm^{-1} whereas $2\nu_2$ is moved up from 1334 to 1388 cm^{-1}. Note that ν_1 and $2\nu_2$ belong to the same irreducible representation.

GROUP FREQUENCIES

A common feature of the IR or Raman spectra of many compounds is that certain groups such as $>CH_2$, $-CH_3$, $>C=O$, etc., have more or less the same frequencies irrespective of the molecular environment. This means that any of these groups must vibrate relatively independently of the rest of the molecule. Thus we can speak of *group frequencies* as a means of identifying their presence.

For a single C—H bond we can have only the stretching and bending vibrations.

$$\overset{\leftarrow}{C}\text{—}\vec{H} \qquad\qquad C\text{—}H \;\Big\rangle$$
stretching   bending

With groups like methylene or methyl the situation is much less straightforward since the protons can move in more complex modes. The problem can again be dealt with group theory.[8] Here we shall merely show the various vibrations of an apparently simple methylene group (Fig. 8-21).[9]

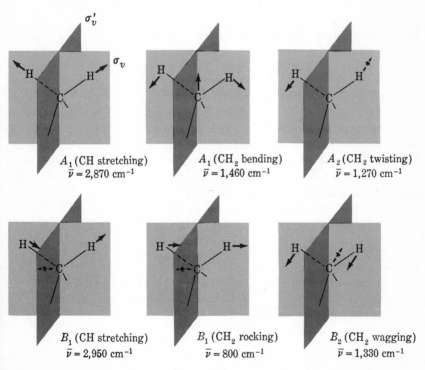

Fig. 8-21 Various vibrational modes of the methylene group. (*By permission of G. M. Barrow, "Introduction to Molecular Spectroscopy," McGraw-Hill Book Company, New York, 1962.*)

8-8 APPLICATIONS

IR SPECTROSCOPY

Chemical analysis   IR is one of the most widely used techniques in chemical analysis.   Its main use lies in the identification of functional groups and molecules rather than the determination of molecular parameters, as in the microwave case.   Since the IR spectrum of any molecule is almost always unique, the matching of the IR spectra of the unknown and known compounds provides the best means of identification.   Similarity between NMR spectra of two compounds does not necessarily mean that they have the same structure since usually only one type of nuclei, that is, the proton, is shown in the spectra.

IR can also be used in quantitative analysis since the absolute concentration can be measured by Beer's law.   It can therefore also be used in kinetics studies to measure the gradual appearance or disappearance of certain lines by recording the spectrum of the reaction mixture at different intervals.   The limitation here is that the reaction must proceed much more slowly than the time required to record the spectrum.

Hydrogen-bonding studies[10]   Consider the following equilibrium due to hydrogen-bond formation

$$X—H + Y \rightleftharpoons X—H \cdots Y$$

where X and Y are groups which contain the electronegative atoms O, N, or F to which the hydrogen is attached.   The X—H stretching frequency will be modified in the hydrogen-bonded complex, and new bands due to the torsional, bending, and stretching motions will appear.   By measuring the intensity of the free and hydrogen-bonded X—H band as a function of the concentration of Y, it is possible to evaluate the equilibrium constant K which is given by

$$K = \frac{[X—H \cdots Y]}{[X—H][Y]}$$

and from the temperature measurement of K we can also obtain the thermodynamic quantities $\Delta G°$, $\Delta H°$, and $\Delta S°$.

As an example let us consider the o-chlorophenol molecule.[11]   This molecule can exist in the cis and trans form; the former is stabilized by the intramolecular hydrogen bonding

The first overtones of the O—H stretching peak are at 6910 cm⁻¹ for the cis and 7050 cm⁻¹ for the trans isomer. At room temperature the integrated intensity of the cis peak is about 10 times that of the trans, showing that the cis form is thermodynamically more stable.

Conformational studies Conformational changes of the type

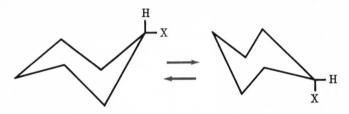

where X = F, Cl, Br, I, —OH, etc., can often be conveniently studied by IR. In general the frequency of the C—X stretching in the equatorial position is higher than that in the axial position; presumably there is less steric interaction with the adjacent protons in the former case. The conformational equilibrium constant K is given by

$$K = \frac{c_e}{c_a}$$

where c_e/c_a is the ratio of the integrated intensities of the C—X stretching peaks in the equatorial and axial positions. For most compounds the free energy difference for the two isomers lies below 1 kcal mole⁻¹.[12]

RAMAN SPECTROSCOPY

The Raman technique has not been used for compound identification in the same way that IR has because of the considerable difficulties in detection as well as the relatively poor resolution. However, with the use of lasers as the exciting source it seems likely that it will soon catch up with IR as a routine analytical tool. There are great advantages to applying both IR and Raman techniques to the same compounds. In fact many of the vibrational spectra can be completely assigned only when both techniques are employed (for example, molecules with a center of symmetry). In general IR is more suitable for molecular vibrations whose frequencies lie above 650 cm⁻¹. This includes most of the organic compounds. On the other hand, Raman can be extended down to as low as 100 cm⁻¹ and is therefore ideally suited for studying weak vibrations such as metal-ligand stretchings, most of which lie below 600 cm⁻¹. There is also the additional advantage that polar solvents such as water can be used in the Raman work. Three examples of applications of Raman spectroscopy will now be given.

Structure of complex ions in solution   One of the early Raman applications is the determination of the structure of complex ions of mercury,
thallium, and silver in solution.[13]   Mercurous ions in aqueous solution
can be either Hg^+ or Hg_2^{++}.   A line is observed which is attributed to the
Hg—Hg stretching, and hence the ions must be in the Hg_2^+ form.   Similar results were also obtained for the thallous (Tl_2^{++}) and argentous
(Ag_2^+) ions.

Ionic equilibria in solution   Consider the dissociation process

$$HNO_3 + H_2O \rightleftharpoons H_3O^+ + NO_3^-$$

By monitoring the intensity of the nitrate ion and the nitric acid in the
Raman spectrum it is possible to obtain the dissociation constant of
HNO_3.[14]   Similarly the dissociations

$$H_2SO_4 + H_2O \rightleftharpoons H_3O^+ + HSO_4^-$$
$$HSO_4^- + H_2O \rightleftharpoons H_3O^+ + SO_4^{--}$$

were also studied.   It was found that the concentration of the SO_4^{--}
ions is small except in very dilute solutions of H_2SO_4.[15]

Nature of bonding   Consider the totally symmetric vibrations of the
tetrahedral complex ML_4 where M represents the metal atom and L the
ligand (for example, $ZnCl_4^{--}$, $CdCl_4^{--}$, $HgCl_4^{--}$, etc.) and the octahedral complex ML_6 (SiF_6^{--}, PF_6^-, SF_6) shown in Fig. 8-22.   Such
vibrations are IR inactive since only the bond lengths change during the
vibration.   However, these same vibrations give rise to intense lines in
the Raman spectra.   From the M—L bond stretching-force constant it
is possible to obtain useful information about the strength of the metal-
ligand bond.[16]   The oxy-anions such as PO_4^{3-}, SO_4^{--}, and ClO_4^{--} have
much larger force constants.   This is taken as evidence that there is
d_π-p_π bonding between the central atom and the oxygen atom in addition
to the σ bonding.

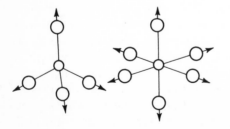

Fig. 8-22 Totally symmetric vibrations of ML_4 and ML_6 complexes.
[*From R. S. Tobias, J. Chem. Educ.,*
44: *40 (1967).   By permission of the
Journal of Chemical Education.*]

PROBLEMS

8-1. For the following symmetric linear molecules calculate the statistical weights which arise for levels with even and odd J values: $C_2{}^{13}D_2{}^2$, $C_3{}^{12}O_2{}^{16}$, $C_3{}^{13}O_2{}^{16}$, and $N_2{}^{14}$, given $I(C^{13}) = \frac{1}{2}$, $I(D^2) = 1$, $I(C^{12}) = 0$, and $I(N^{14}) = 1$.

8-2. Verify the force constants in Table 8-1 for the molecules H_2, HCl, and I_2 using the fundamental frequencies in the same table.

8-3. Acetylacetone ($CH_3COCH_2COCH_3$) shows the normal aliphatic ketone band at 1709 cm^{-1}. There is a broadband between 1640 and 1540 cm^{-1} which is attributed to the O—H stretching. What is the origin of this band?

8-4. Diimide (N_2H_2) can exist in both the cis and trans form. Determine the following properties for these two isomers:

    (a) The symmetry point groups to which they belong.

    (b) The number of fundamentals and the symmetry species to which they belong.

    (c) The IR and Raman activities of these fundamentals.

8-5. The normal modes of vibration of a planar molecule such as BCl_3 can be represented by

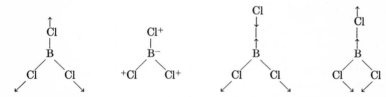

Determine the symmetry of the vibrational modes and the spectral activity of each mode.

8-6. The normal modes of vibration of a pyramidal molecule PH_3 can be represented by

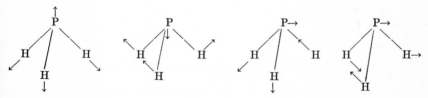

Determine the symmetry of the vibrational modes and the spectral activity of each mode.

8-7. The following IR bands are attributed to C—H stretching in C_2H_4: 2991, 3010, 3107, and 3109 cm^{-1}. How would you confirm this assignment?

SUGGESTIONS FOR FURTHER READING

Introductory

Barrow, G. M.: "The Structure of Molecules," W. A. Benjamin, Inc., New York, 1963.

Dunford, H. B.: "Elements of Diatomic Molecular Spectra," Addison-Wesley Publishing Company, Inc., Reading, Mass., 1968.

Intermediate-advanced

Barrow, G. M.: "Introduction to Molecular Spectroscopy," McGraw-Hill Book Company, New York, 1963.

Colthup, N. B., L. H. Daly, and S. E. Wiberley: "Introduction to Infrared and Raman Spectroscopy," Academic Press, Inc., New York, 1964.

Cotton, F. A.: "Chemical Applications of Group Theory," chap. 9, Interscience Publishers, a division of John Wiley & Sons, Inc., New York, 1963.

Herzberg, G.: "Molecular Spectra and Molecular Structure. I. Spectra of Diatomic Molecules," D. Van Nostrand Company, Inc., Princeton, N.J., 1950.

Herzberg, G.: "Molecular Spectra and Molecular Structure. II. Infrared and Raman Spectra of Polyatomic Molecules," D. Van Nostrand Company, Inc., Princeton, N.J., 1945.

Szymanski, H. A. (ed.): "Raman Spectroscopy," Plenum Press, Plenum Publishing Corporation, New York, 1967.

West, W. (ed.): Chemical Applications of Spectroscopy, in "Techniques of Organic Chemistry," vol. 9, Interscience Publishers, Inc., New York, 1956.

READING ASSIGNMENTS

Near Infrared Spectra, O. H. Wheeler, *J. Chem. Educ.*, **37**: 234 (1960).

Infrared Spectroscopy: A Chemist's Tool, *J. Chem. Educ.*, **37**: 651 (1960).

Inorganic Infrared Spectroscopy, J. R. Ferraro, *J. Chem. Educ.*, **38**: 201 (1961).

Vibration-Rotation Spectrum of HCl, F. E. Stafford, C. W. Holt, and G. L. Paulson, *J. Chem. Educ.*, **40**: 245 (1963).

Molecular Structure and Thermodynamics Properties of HCN and DCN, R. Little, *J. Chem. Educ.*, **43**: 2 (1966).

The Infrared Spectra of C_2H_2 and C_2D_2, L. W. Richards, *J. Chem. Educ.*, **43**: 644 (1966).

Raman Spectroscopy in Inorganic Chemistry. I. Theory, R. S. Tobias, *J. Chem. Educ.*, **44**: 2 (1967).

Raman Spectroscopy in Inorganic Chemistry. II. Applications, R. S. Tobias, *J. Chem. Educ.*, **44**: 70 (1967).

"C—H Bond Strengths," G. J. Boobyer and A. P. Cox, *J. Chem. Educ.*, **45**: 18 (1968).

Recent Trends and Developments in Inorganic Far Infrared Spectroscopy, J. R. Ferraro, *Anal. Chem.*, **40**: 24A (1968).

Infrared Spectroscopic Studies of Hindered Internal Rotation, A. J. Woodward and N. Jonathan, *J. Chem. Educ.*, **46**: 756 (1969).

Infrared Determination of Stereochemistry in Metal Complexes, M. Y. Darensbourg and D. J. Darensbourg, *J. Chem. Educ.*, **47**: 33 (1970).

Infrared Spectrometry of Inorganic Salts, M. N. Ackermann, *J. Chem. Educ.*, **47**: 69 (1970).

REFERENCES

1. D. F. Eggers, Jr., N. W. Gregory, G. D. Halsey, Jr., and B. S. Rabinovitch, "Physical Chemistry," p. 160, John Wiley & Sons, Inc., New York, 1964.
2. P. M. Morse, *Phys. Rev.*, **34**: 57 (1929).
3. L. Pauling and E. B. Wilson, Jr., "Introduction to Quantum Mechanics," McGraw-Hill Book Company, New York, 1935.

4. H. Margenau and G. M. Murphy, "The Mathematics of Physics and Chemistry," p. 161, D. Van Nostrand Company, Inc., Princeton, N.J., 1959.

5. G. Herzberg, "Molecular Spectra and Molecular Structure. II. Infrared and Raman Spectra of Polyatomic Molecules," p. 26, D. Van Nostrand Company, Inc., Princeton, N.J., 1966.

6. *Ibid.*, p. 239.

7. *Ibid.*, p. 215.

8. G. M. Barrow, "Introduction to Molecular Spectroscopy," p. 198, McGraw-Hill Book Company, New York, 1963.

9. *Ibid.*, p. 200.

10. G. C. Pimental and A. L. McClellan, "The Hydrogen Bond," W. H. Freeman and Company, San Francisco, 1960.

11. L. Pauling, "The Nature of the Chemical Bond," p. 485, Cornell University Press, 1960.

12. J. A. Hirsch, in N. L. Allinger and E. L. Eliel (eds.), "Topics in Stereochemistry," vol. 1, p. 199, Interscience Publishers, a division of John Wiley & Sons, Inc., New York, 1967.

13. L. A. Woodward, *Phil. Mag.*, **18**: 823 (1934).   D. N. Waters and L. A. Woodward, *J. Chem. Soc.*, 3250 (1954).

14. O. Redlich and J. Bigeleisen, *J. Am. Chem. Soc.*, **65**: 1883 (1943).

15. C. K. Ingold, D. J. Millen, and H. G. Poole, *J. Chem. Soc.*, 2576 (1950).   D. J. Millen, *J. Chem. Soc.*, 2589, 2600, 2606 (1950).   C. K. Ingold and D. J. Millen, *J. Chem. Soc.*, 2612 (1950).

16. L. A. Woodward, *Trans. Faraday Soc.*, **54**: 1271 (1958).

9
Electronic Spectroscopy of Atoms

9-1 INTRODUCTION

It was well known in the nineteenth century that both the absorption and emission spectrum of atomic hydrogen consisted of a series of sharp, well-defined lines.   In 1885 Balmer proposed the following formula to account for the regularity of the lines in the spectra

$$\bar{\nu} = R_{\mathrm{H}} \left(\frac{1}{n''^2} - \frac{1}{n'^2} \right) \tag{9-1}$$

where $\bar{\nu}$ is the wave number of a particular line, R_{H} is a constant called Rydberg constant, and n' and n'' are integral numbers.   Here we have $n'' < n'$.   The following series were named after their discoverers:

Series	n''	n'
Lyman	1	2, 3, . . .
Balmer	2	3, 4, . . .
Paschen	3	4, 5, . . .
Brackett	4	5, 6, . . .
Pfund	5	6, 7, . . .

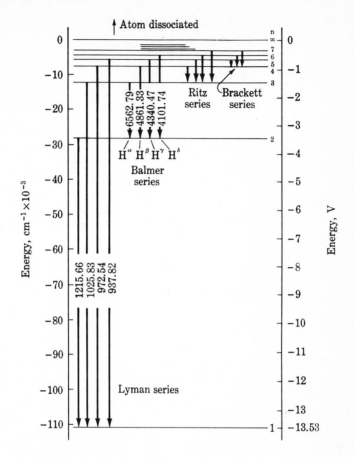

Fig. 9-1 Energy levels and the various series for the atomic hydrogen spectrum. (*Jeff C. Davis, Jr., "Advanced Physical Chemistry—Molecules, Structure, and Spectra."* Copyright © 1965, The Ronald Press Company, New York.)

Figure 9-1 shows the energy levels and the various series for the atomic hydrogen spectrum.

In 1913 Bohr stated that electrons in atoms can occupy only discrete energy levels and the lines in the atomic spectra can be accounted for by the equation

$$\Delta E = E_2 - E_1 = h\nu \tag{9-2}$$

where E_2 and E_1 are the energy levels. From this Bohr was able to derive the Rydberg constant as

$$R_{\mathrm{H}} = \frac{2\pi^2 m_e e^4}{ch^3} \qquad \mathrm{cm}^{-1} \tag{9-3}$$

The solution of the Schrödinger equation of hydrogen atom gives the energy E_n [Eq. (1-41)]

$$E_n = -\frac{2\pi^2\mu e^4}{n^2 ch^3} \quad cm^{-1} \tag{9-4}$$

From Eq. (9-4) it is clear that n' and n'' in Eq. (9-1) are simply the principal quantum numbers.

The spectra of hydrogenlike ions such as He^+, Li^{++}, . . . , are also well accounted for by Eq. (9-4) which is now given by

$$E_n = -\frac{2\pi^2 Z^2 \mu e^4}{n^2 ch^3}$$

where Z is the atomic number.

The probability of transition from state m to n is proportional to μ_{nm}^2 where μ_{nm} is given by

$$\int \psi_n^* \mathbf{u} \psi_m \, d\tau$$

where $\mathbf{u}$ is the dipole-moment operator defined as

$$\mathbf{u} = \sum_{i=1}^{n} e\mathbf{r}_i$$

where e is the electronic charge and $\mathbf{r}_i$ is the distance of the ith electron from the nucleus. The summation extends over all n electrons. The selection rules are

$$\Delta n = \pm 1, \pm 2, \pm 3, . . .$$
$$\Delta l = \pm 1$$

9-2 EXPERIMENTAL TECHNIQUES

In the emission studies the sample is excited by flame, electric arc, or plasma discharge. The emitted radiation passes through either a monochromator or a spectrograph and is then recorded photographically. The absorption technique is also useful, especially when the transitions associated with the ground state are studied. Figure 9-2 shows the schematic diagrams for absorption and emission studies.

A third type of atomic spectroscopy is called atomic fluorescence spectrophotometry. In contrast to the usual emission techniques which use thermal excitation, the excitation in atomic fluorescence spectrophotometry is caused by electromagnetic radiation. It was first observed by Wood in 1904 that when the yellow light emitted by sodium atoms in a flame was focused on a bulb containing metallic sodium at low pressure, the sodium vapor in the bulb absorbed the radiation and gave off a faint

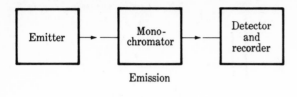

Emission

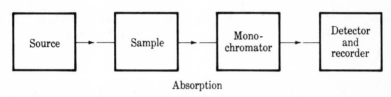

Absorption

Fig. 9-2 Schematic diagrams for absorption and emission studies.

yellow glow. He concluded that the emitted radiation is resonantly absorbed.† Atomic fluorescence spectrophotometry is very useful in analytical chemistry, but we shall not discuss it further in this text.

9-3 THEORY

HYDROGEN AND HYDROGENLIKE ION SPECTRA

In Sec. 9-1 only the orbital angular momentum l of the hydrogen atom was considered. If we consider the spin angular momentum as well, the resultant angular momentum j is given by

$$\mathbf{j} = \mathbf{l} + \mathbf{s} \tag{9-5}$$

where

$$j = |l + s|, |l + s - 1|, \ldots, |l - s|$$

Due to the coupling between the spin and the orbital motion most of the energy levels shown in Fig. 9-1 will be split into doublets. For example, for the $2p$ case we have $l = 1$ and $s = \frac{1}{2}$, and hence $j = \frac{3}{2}$ and $\frac{1}{2}$. These states are labeled by the symbols $^{2s+1}L_j$ so that we have $^2P_{\frac{3}{2}}$ and $^2P_{\frac{1}{2}}$. On the other hand the $1s$ orbital remains a singlet since $l = 0$, that is, $^2S_{\frac{1}{2}}$. Thus the $1s \rightarrow 2p$ transition will give rise to two lines shown in Fig. 9-3.

† This phenomenon is exactly analogous to the nuclear fluorescence experiment discussed in Chap. 5.

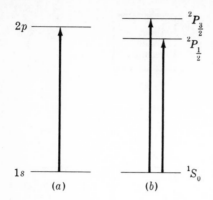

Fig. 9-3 (*a*) No spin-orbit coupling
(*b*) Spin-orbit coupling present.

This coupling is very weak for hydrogen and the first few hydrogen like ions; the doublets and compound doublets can be resolved only with spectrometers of the highest resolving power.

ALKALI METAL SPECTRA

These metals have a single electron in the outer s orbital, and their spectra bear a superficial resemblance to the hydrogen spectrum. Figure 9-- shows the term diagram of Li. Contrary to the hydrogen case the fine structure in the alkali metal spectra can be easily resolved. For example, the doublet from $3^2P_{\frac{3}{2}}$ and $3^2P_{\frac{1}{2}}$ to $3^2S_{\frac{1}{2}}$ of sodium appear at 5889.9 and 5895.9 Å. These are the well-known sodium D lines.

COUPLING OF ORBITAL AND SPIN ANGULAR MOMENTA

When there is more than one valence electron present the situation becomes more complex. There are two ways in which the orbital angular momentum can be coupled to the spin angular momentum.

1. *Russell-Saunders coupling* (see Appendix 6) In this coupling scheme the individual orbital and spin angular momentum are added up separately and the two sums are then combined to give the resultant angular momentum J. We write

$$l_1 + l_2 + \cdots = \sum_i^n l_i = L$$

$$s_1 + s_2 + \cdots = \sum_i^n s_i = S$$

and finally

$$\mathbf{J} = \mathbf{L} + \mathbf{S} \tag{9-6}$$

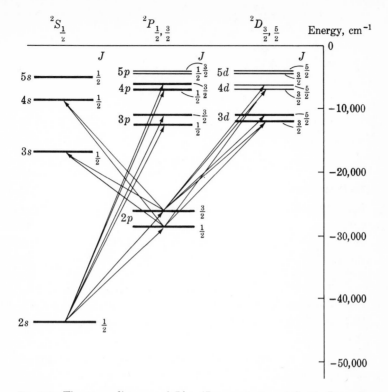

Fig. 9-4 The term diagram of Li. (*By permission of C. N. Banwell, "Fundamentals of Molecular Spectroscopy," McGraw-Hill Publishing Company, Ltd., London, 1966.*)

The possible values of the quantum numbers are

$$L = (l_1 + l_2 + \cdots), (l_1 + l_2 + \cdots -1), \ldots, 0$$
$$S = (s_1 + s_2 + \cdots), (s_1 + s_2 + \cdots -1), \ldots, 0 \text{ (or } \tfrac{1}{2})$$

and

$$J = L + S, L + S - 1, \ldots, |L - S|$$

where L and S are the maximum values.   This coupling scheme applies when there is strong coupling among the orbital angular momenta (and the spin angular momenta) themselves.   It is a good approximation for light atoms (up to the alkaline earth metals).

2. *j-j coupling*   This is opposite to the Russell-Saunders case; that is, there is strong coupling between the individual orbital and spin angular

momentum. We write

$$l_1 + s_1 = j_1$$
$$l_2 + s_2 = j_2$$
$$\cdot \; \cdot \; \cdot \; \cdot \; \cdot \; \cdot \; \cdot$$

and

$$j_1 + j_2 + \cdots = \sum_i^n j_i = J$$

This coupling scheme applies mainly to the heavy atoms.

TERM SYMBOLS

Let us consider the ground-state helium atom ($1s^2 2s^2$). Its terms can be evaluated using the Russell-Saunders coupling scheme. The two electrons are paired and hence the only term is $^1S_0$. The situation is more complex when we consider the carbon atom ($1s^2 2s^2 2p^2$). There are many ways to label these two electrons and it is convenient to summarize them as in Table 9-1. We note that the substates are formed in accordance with the Pauli exclusion principle. The 15 substates give rise to only five terms as follows: One term must contain the maximum value of M_L, which is 2. We have

$$M_L = 2, 1, 0, -1, -2 \qquad (1), (3), (7), (12), \text{ and } (15)$$
$$M_S = 0$$

Table 9-1 Substates for two equivalent p electrons

m_{l_1}	m_{l_2}	M_L	m_{s_1}	m_{s_2}	M_S	Substate
1	1	2	$+\frac{1}{2}$	$-\frac{1}{2}$	0	(1)
1	0	1	$\frac{1}{2}$	$\frac{1}{2}$	1	(2)
1	0	1	$\frac{1}{2}$	$-\frac{1}{2}$	0	(3)
1	0	1	$-\frac{1}{2}$	$\frac{1}{2}$	0	(4)
1	0	1	$-\frac{1}{2}$	$-\frac{1}{2}$	-1	(5)
1	-1	0	$\frac{1}{2}$	$\frac{1}{2}$	1	(6)
1	-1	0	$\frac{1}{2}$	$-\frac{1}{2}$	0	(7)
1	-1	0	$-\frac{1}{2}$	$\frac{1}{2}$	0	(8)
1	-1	0	$-\frac{1}{2}$	$-\frac{1}{2}$	-1	(9)
0	0	0	$\frac{1}{2}$	$-\frac{1}{2}$	0	(10)
0	-1	-1	$\frac{1}{2}$	$\frac{1}{2}$	1	(11)
0	-1	-1	$\frac{1}{2}$	$-\frac{1}{2}$	0	(12)
0	-1	-1	$-\frac{1}{2}$	$\frac{1}{2}$	0	(13)
0	-1	-1	$-\frac{1}{2}$	$-\frac{1}{2}$	-1	(14)
-1	-1	-2	$\frac{1}{2}$	$-\frac{1}{2}$	0	(15)

hich gives the $^1D_2$ term.   Next we choose the $M_L = 1$ value.   We have

$M_L = 1, 0, -1$ (2), (6), and (11)
$M_S = 1$

$M_L = 1, 0, -1$ (4), (7), and (13)
$M_S = 0$

$M_L = 1, 0, -1$ (5), (9), and (14)
$M_S = -1$

hich give the $^3P_2$, $^3P_1$, and $^3P_0$ terms (since $J = L + S, L + S - 1$, ad $L - S$ and here we have $J = 2, 1$, and 0).   Finally, the last substate is

$M_L = 0$ (10)
$M_S = 0$

hich gives the $^1S_0$ term.

UND'S RULE OF MULTIPLICITY

iven the terms of any atom our next task is to determine the relative ·der of energies.   This is done by Hund's rule: (1) When two states ave the same L value, the one having the higher S value lies lower in ıergy.   (2) When two states have the same S value, the one having igher L value lies lower in energy.   (3) For orbitals that are less than alf-filled, states having lower J values lie lower in energy; for orbitals ıat are more than half-filled, states having higher J values lie lower in ıergy.   Applying Hund's rule to the carbon atom we obtain the follow- ıg states in the order of increasing energy

$$^3P_2 < {}^3P_1 < {}^3P_0 < {}^1D_2 < {}^1S_0$$

HE ZEEMAN EFFECT

n the presence of a magnetic field each line is split into a number of com- onents; the splitting is proportional to the strength of the magnetic eld.   This effect, first discovered by Zeeman in 1896, is due to the inter- ction between the external magnetic field and the magnetic moment ssociated with the electron spin and orbital motion.   If the magnetic ıoment is only due to the orbital motion, that is, for the singlet state $S = 0$), we have $L = J$ and

$$\mu = -\frac{e\hbar}{2m_ec} \sqrt{L(L + 1)} = -\frac{e\hbar}{2m_ec} \sqrt{J(J + 1)}$$

$$\simeq -\frac{e\hbar}{2m_ec} J \qquad (9\text{-}7)$$

The component of the magnetic moment in the direction of the magnetic field μ_H is given by

$$\mu_H = -\frac{e\hbar}{2m_e c} M_J \tag{9-8}$$

where M_J is the component of J in the direction of the field. The energy of this magnetic interaction E_{int} is given by

$$E_{int} = -\mathbf{\mu}_H \cdot \mathbf{H} \tag{9-9}$$

where H is the applied field. Thus

$$E_{int} = \beta H M_J \tag{9-10}$$

Let us consider a specific transition, say $^1P_1 \to {}^1S_0$, in the absence and presence of the magnetic field (Fig. 9-5). The selection rules are $\Delta M_J = 0, \pm 1$.†

The above case involves only the singlets and is known as the *normal* Zeeman effect. The *anomalous* Zeeman effect occurs when we are dealing with states of higher spin multiplicities. This is due to the added contribution to the magnetic moment by the spin motion of the electron. We have, in the Russell-Saunders coupling scheme,

$$\mathbf{J} = \mathbf{L} + \mathbf{S}$$
$$\mu_J = -g_J \beta \hbar \sqrt{J(J + 1)}$$
$$\simeq -g_J \beta \hbar J \tag{9-11}$$

† For $\Delta J = 0$, $M_J = 0 \to 0$ transitions are forbidden. The selection rules are the same as that for the normal Zeeman effect.

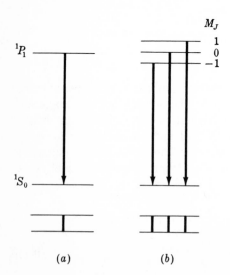

Fig. 9-5   (a) No magnetic field present
(b) Magnetic field present.

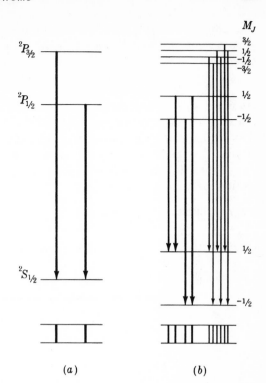

Fig. 9-6 (*a*) No magnetic field present. (*b*) Magnetic field present.

(*a*)   (*b*)

where g_J is the Landé g factor (see Appendix 6).   The energy of interaction is given by

$$E_{\text{int}} = g_J \beta H M_J \qquad (9\text{-}12)$$

Figure 9-6 shows the anomalous Zeeman effect for the sodium D lines.

THE PASCHEN–BACK EFFECT

When the strength of the applied magnetic field becomes greater than the multiplet splittings, we have the so-called Paschen-Back effect.   In this region the orbital and spin angular momenta are no longer coupled to each other but are coupled separately to the magnetic field.   In this case only the normal Zeeman effect is observed even when doublets or higher spin states are present.   It is important to note that although in principle ESR and NMR can be observed at any magnetic field the usual experiments are always performed in the Paschen-Back region.

THE STARK EFFECT

In the presence of an electric field ϵ the atomic levels will also be split and complex spectra result.   This effect was first observed by Stark in 1913.

The interaction here is between ϵ and the *induced* electric dipole moment μ_ϵ as a result of electron polarization. The energy of interaction E_{int} is given by

$$E_{int} = \boldsymbol{\mu}_\epsilon \cdot \boldsymbol{\epsilon} \tag{9-13}$$

However, since μ_ϵ is proportional to ϵ, the splitting of the levels is proportional to ϵ^2. The differences between the Zeeman and Stark effects are that in the latter case the splitting pattern is not symmetric about the original line, and the energy values depend on $M_J{}^2\epsilon^2$.

Figure 9-7 summarizes the various interactions discussed in this section for a many-electron atom.

X-RAY FLUORESCENCE SPECTROSCOPY

Another branch of atomic spectroscopy which needs some mention involves the excitation of the inner electrons. Since this normally

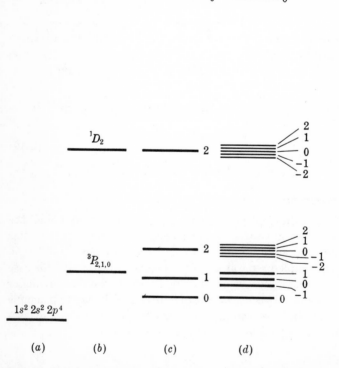

Fig. 9-7 Diagram summarizing the various interactions. (*a*) Electronic configuration. (*b*) Terms as a result of electron repulsion. (*c*) Levels as a result of spin-orbit coupling. (*d*) States as a result of external magnetic field.

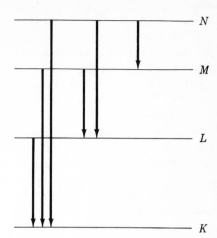

Fig. 9-8 The K, L, and M series.

requires much higher energy, the frequency of excitation is in the x-ray region.

When an electron originally in the K shell is excited to a higher vacant level, one of the outer electrons immediately falls into the vacancy in the K shell, emitting radiation in the process. In this way a number of lines will be observed which form the K series. Similarly we can have the L, M, N, . . . series as shown in Fig. 9-8.

The quantitative aspect of these x-ray spectra was studied by Moseley, who found that the frequency ν of the line belonging to any particular series was related to the atomic number Z of the element according to the equation

$$\sqrt{\nu} = a(Z - \sigma) \tag{9-14}$$

where a and σ are constants.

9-4 APPLICATIONS

Chemical analysis Since each element has its own characteristic emission spectrum, the obvious application of atomic emission spectroscopy is in the qualitative and quantitative chemical analyses. In the latter case the intensity of the lines is compared with the standard, and the concentration can be readily evaluated. The emission spectra of most metallic elements fall within the region of 2000 to 6000 Å. Nonmetallic elements present some practical difficulties since they are not electrically conducting and therefore cannot be made into an electrode in a discharge.

Atomic absorption as an analytic tool was developed more recently (since 1950). To utilize atomic absorption spectroscopy a sample must be converted into the atomic vapor. It is of interest to point out here that there are no examples in which the absorption lines of one element overlap with another and hence the technique is almost free of spectral interferences (molecular bands sometimes interfere slightly). Of almost all the work done in this branch, an open flame (for example, air and acetylene) is used to produce the atomic vapor from the sample. The requirement is that the metal atoms must be in the ground state, that is, they must be neither excited nor ionized. Metals whose compounds are not decomposed by the flame or those which form oxides or hydroxides when heated give problems of greatly reduced sensitivity.

The atomic absorption spectroscopy technique is also widely used in biochemistry. For example, Ca, Mg, Na, and K contents in blood and urine can be easily determined.

PROBLEMS

9–1. The ground state electronic configuration of an atom is (filled orbitals) np^3. Derive the term symbols and draw an energy level diagram in accordance with Hund's rule.

9–2. Describe the procedure for obtaining the ionization potential of atoms from their spectra. What is the ionization potential of hydrogen in electronvolts?

9–3. Suggest an experiment to verify the Russell-Saunders coupling in light atoms.

SUGGESTIONS FOR FURTHER READING

Introductory

Hochstrasser, R. M.: "Behavior of Electrons in Atoms," W. A. Benjamin, Inc., New York, 1965.

Intermediate-advanced

Herzberg, G.: "Atomic Spectra and Atomic Structure," Dover Publications, Inc., New York, 1944.

Mitchell, A. C. G., and M. W. Zemansky: "Resonance Radiation and Excited Atoms," Cambridge University Press, New York, 1961.

Kuhn, H. G.: "Atomic Spectra," Academic Press, Inc., New York, 1962.

READING ASSIGNMENTS

Atomic Spectra, F. E. Stafford and J. H. Wortman, *J. Chem. Educ.*, **39**: 630 (1962).

The Spectrum of Atomic Hydrogen, J. L. Hollenberg, *J. Chem. Educ.*, **43**: 216 (1966).

10
Electronic Spectroscopy of Diatomic Molecules

10-1 INTRODUCTION

The electronic spectra of diatomic molecules are very complex in general, the reason being that in addition to the electronic transitions there are almost always simultaneous rotational and vibrational changes. Born and Oppenheimer showed that the approximate solution of the Schrödinger wave equation for a molecule can be obtained if we assume that electronic and nuclear motions are completely independent of each other (see Appendix 7). The result of this approximation is that we can express the total energy of the molecule E_T as

$$E_T = E_e + E_n$$

where E_n is given by

$$E_n = E_{\text{trans}} + E_{\text{rot}} + E_{\text{vib}}$$

Thus†

$$E_T = E_e + E_{\text{rot}} + E_{\text{vib}} \tag{10-1}$$

Note that Eq. (10-1) is the same as Eq. (1-85).   Since

$$E_{\text{rot}} = B_v J(J + 1) - D_v [J(J + 1)]^2$$

and

$$E_{\text{vib}} = (v + \tfrac{1}{2})\bar{\nu}_e - x(v + \tfrac{1}{2})^2 \bar{\nu}_e$$

the energy change (in cm^{-1}) in an electronic transition ΔE_{evr} is given by‡

$$\Delta E_{evr} = (E'_e - E''_e) + [(v' + \tfrac{1}{2})\bar{\nu}'_e - (v'' + \tfrac{1}{2})\bar{\nu}''_e]$$
$$+ [B'_v J'(J' + 1) - B''_v J''(J'' + 1)] \tag{10-2}$$

In an electronic transition there are no selection rules regarding the change in the vibrational quantum number v.   As we shall see later that although certain transitions are preferred, Δv may be positive, negative, or zero.   In diatomic spectra we use the terminology *band* to describe a definite pair of quantum numbers v' and v'' and *band system* to describe an ensemble of bands associated with the same electronic transition.   The term 0-0 *band* applies to a specific electronic transition in which both v' and v'' are zero.   The term *hot band* applies to the transition from the ground electronic but vibrationally excited state to a higher electronic but lower vibrational state, that is, $v' < v''$.   A series of bands having the same $(v'\text{-}v'')$ value is called a *sequence*, and the term *progression* applies to a set of bands having the same value of either v' or v''.

10-2 THEORY

MOLECULAR ORBITAL TREATMENT OF DIATOMIC MOLECULES

An understanding of the electronic spectra requires a knowledge of the number of and the spacing between the electronic levels.   There are two approaches to this problem.   Let us consider a homodiatomic molecule A_2.   We start with two extreme situations: (1) $r = 0$, and (2) $r \rightarrow \infty$, where r is the internuclear distance; we will then try to obtain the approximate solutions for the real system.

The united atom model   In this extreme case we have a many-electron atom in general.   The Schrödinger wave equation can be approximately solved to give the energy levels shown in region a of Fig. 10-1.   Just as in the atomic case, each orbital is characterized by the total angular

---

† We omit here the translational energy which is not quantized in our subsequent discussion.

‡ We assume that the molecule behaves like a rigid rotator and a harmonic oscillator.

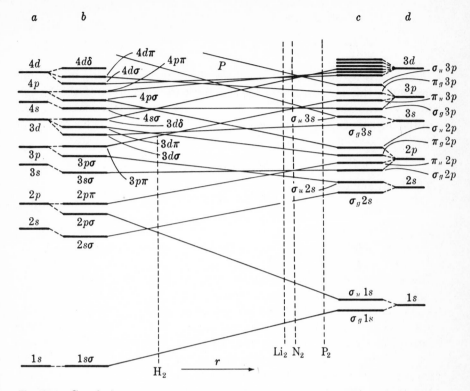

Fig. 10-1 Correlation of molecular orbitals for homonuclear diatomic molecules. [*From "Molecular Spectra and Molecular Structure: I, Spectra of Diatomic Molecules," by Gerhard Herzberg, 2d ed. Copyright ©, by Litton Educational Publishing, Inc.; by permission of Van Nostrand Reinhold Company.*]

momentum, that is, $L = 0, 1, 2, \ldots$. Now suppose this united atom is pulled apart to form a diatomic molecule; there will then be an electrostatic field in the direction of the internuclear axis. This means that the electric field now has an axial rather than the spherical symmetry as in the united atom case. Hence it will be necessary to characterize the orbitals by the components of L along the molecular axis, which is represented by Λ. We have

$$\Lambda = 0, 1, 2, \ldots, L \qquad (10\text{-}3)$$

In the atomic case we have

$$L = 0, 1, 2, \ldots$$
$$S \ P \ D \ \cdots$$

and in the molecular case

$$\Lambda = 0, 1, 2, \ldots$$
$$\Sigma \quad \Pi \quad \Delta \quad \cdots$$

In atoms, S is the total spin of all the electrons present, and in diatomic molecules this is represented by Σ. With the exception of the Σ state ($\Lambda = 0$), the orbital motion of the electrons produces a magnetic field in the direction of the internuclear axis. Thus for every value of S there are $2S + 1$ values of Σ given by

$$\Sigma = S, S - 1, \ldots, 0, \ldots, -(S - 1), -S \qquad (10\text{-}4)$$

The total angular momentum Ω is given by†

$$\Omega = \Lambda + \Sigma \qquad (10\text{-}5)$$

Region b in Fig. 10-1 shows the splittings of the energy levels when the "nucleus" of the united atom is pulled apart to form two atoms.

The separate atom model   In the other extreme case we consider the approach of two atoms from a large distance to form a molecule. The atomic energy levels are shown in region d in Fig. 10-1. As the two atoms are close enough to interact (region c in Fig. 10-1), a molecular orbital is formed in which the two electrons reside, assuming each atom donates one bonding electron. As a first approximation the molecular orbital ψ can be written as a linear combination of the atomic orbitals, that is, in the LCAO approximation we have

$$\psi = \phi_A{}^1 \pm \phi_A{}^2 \qquad (10\text{-}6)$$

where ϕ_A is the atomic orbital and superscripts 1 and 2 indicate the two atoms. The sum of the two atomic orbitals gives the bonding molecular orbital ψ_g

$$\psi_g = \phi_A{}^1 + \phi_A{}^2$$

and the difference gives the antibonding molecular orbital ψ_u

$$\psi_u = \phi_A{}^1 - \phi_A{}^2$$

where g and u indicate whether ψ is symmetric or antisymmetric with respect to the inversion about the center of symmetry. Figure 10-2 shows the potential energy curve of H_2.

For a heteronuclear diatomic molecule AB we have

$$\psi = \phi_A \pm \gamma \phi_B \qquad (10\text{-}7)$$

where γ is a measure of the polarity of the chemical bond.

Figure 10-3 shows the energy correlation diagram for a hetero-diatomic molecule.

† Compare Eq. (10-5) with $\mathbf{J} = \mathbf{L} + \mathbf{S}$.

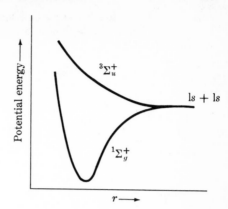

Fig. 10-2 The potential energy curve or H_2.

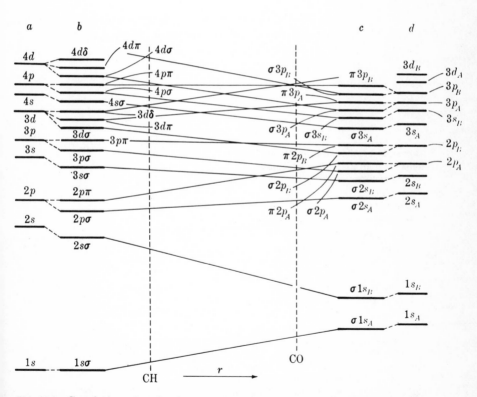

Fig. 10-3 Correlation of molecular orbitals for heteronuclear diatomic molecules. [*From "Molecular Spectra and Molecular Structure: I, Spectra of Diatomic Molecules," by Gerhard Herzberg, 2d ed. Copyright ⓒ, by Litton Educational Publishing, Inc., by permission of Van Nostrand Reinhold Company.*]

Table 10-1   Some examples of the descriptions used for electronic states

K is used to denote two electrons in $1s$ orbits; L is used to denote eight electrons in the $2s$ and $2p$ orbits.

Molecule	Ground configuration		First excited configuration	
H_2	$(\sigma_g 1s)^2$	$^1\Sigma_g{}^+$	$(\sigma_g 1s)\sigma_u 1s$	$^1\Sigma_u{}^+,\ ^3\Sigma_u{}^+$
Li_2	$KK(\sigma_g 2s)^2$	$^1\Sigma_g{}^+$	$KK(\sigma_g 2s)\sigma_u 2s$	$^1\Sigma_u{}^+,\ ^3\Sigma_u{}^+$
N_2	$KK(\sigma_g 2s)^2(\sigma_u 2s)^2(\pi_u 2p)^4$ $(\sigma_g 2p)^2$	$^1\Sigma_g{}^+$	$KK(\sigma_g 2s)^2(\sigma_u 2s)^2(\pi_u 2p)^4$ $(\sigma_g 2p)^1\pi_g 2p$	$^1\Pi_g,\ ^1\Pi_u$
LiH	$K(2s\sigma)^2$	$^1\Sigma^+$	$K(2s\sigma)2p\sigma$	$^1\Sigma^+,\ ^3\Sigma^+$
CH	$K(2s\sigma)^2(2s\sigma)^2 2p\pi$	$^2\Pi$	$K(2s\sigma)^2 2p\sigma(2p\pi)^2$	$^4\Sigma^-,\ ^2\Delta,$ $^2\Sigma^+,\ ^2\Sigma^-$

SOURCE: By permission of G. M. Barrow, "Introduction to Molecular Spectroscopy," McGraw-Hill Book Company, New York, 1962.

The spectroscopic notation for any diatomic molecule, regardless whether in the united-atom or separate-atom model, is given by

$$^{2\Sigma+1}\Lambda$$

Table 10-1 gives some examples of diatomic molecules.

Table 10-2   Allowed electronic transitions for light diatomic molecules

Unequal nuclear charge	Equal nuclear charge
$\Sigma^+ \leftrightarrow \Sigma^+$	$\Sigma_g{}^+ \leftrightarrow \Sigma_u{}^+$
$\Sigma^- \leftrightarrow \Sigma^-$	$\Sigma_g{}^+ \leftrightarrow \Sigma_u{}^-$
$\Pi \leftrightarrow \Sigma^+$	$\Pi_g \leftrightarrow \Sigma_u{}^+,\ \Pi_u \leftrightarrow \Sigma_g{}^+$
$\Pi \leftrightarrow \Sigma^-$	$\Pi_g \leftrightarrow \Sigma_u{}^-,\ \Pi_u \leftrightarrow \Sigma_g{}^-$
$\Pi \leftrightarrow \Pi$	$\Pi_g \leftrightarrow \Pi_u$
$\Pi \leftrightarrow \Delta$	$\Pi_g \leftrightarrow \Delta_u,\ \Pi_u \leftrightarrow \Delta_g$
$\Delta \leftrightarrow \Delta$	$\Delta_g \leftrightarrow \Delta_u$
. . . .	. . . .

SOURCE: From G. Herzberg, "Molecular Spectra and Molecular Structure, I. Spectra of Diatomic Molecules," 2d ed., Copyright ©, by Litton Educational Publishing, Inc. Reproduced by permission of Van Nostrand Reinhold Company.

SELECTION RULES

The selection rules are given by

$$\Delta\Lambda = 0, \pm 1 \qquad \Delta\Sigma = 0$$

Table 10-2 summarizes the allowed electronic transitions for diatomic molecules.

All the above transitions are possible as singlets, doublets, triplets, and so on, but intercombinations—singlet-triplet and similar transitions—are forbidden.

THE FRANCK–CONDON PRINCIPLE

When an electronic transition is accompanied by a simultaneous change in the vibrational quantum number, the Franck-Condon principle can help us to determine the intensity of the bands in the molecular spectra.[1] This principle states that since the time required for a molecule to execute a vibration (about 10^{-12} sec) is much longer than that required for an electronic transition (about 10^{-15} sec), the nuclei do not appreciably alter their positions or kinetic energies during an electronic transition. This means that the internuclear distance r remains unchanged during the transition, and the intensity of the bands can be predicted from the vertical lines as shown in Fig. 10-4. The transition dipole moment μ_{nm} is given by

$$\mu_{nm} = \int \psi_e' \psi_{\text{vib}}' \psi_{\text{rot}}' \mu \psi_e'' \psi_{\text{vib}}'' \psi_{\text{rot}}'' \, d\tau \tag{10-8}$$

The integrals involving ψ_{rot} can be separated out as shown in Sec. 8-3. The dipole moment can be written as a sum of two terms

$$\mu = \mu_e + \mu_n \tag{10-9}$$

where μ_e and μ_n are the electronic and nuclear dipole moments. Thus Eq. (10-8) can be rewritten as

$$\mu_{nm} = \int \psi_e' \mu_e \psi_e'' \, d\tau_e \int \psi_{\text{vib}}' \psi_{\text{vib}}'' \, d\tau_n$$
$$+ \int \psi_e' \psi_e'' \, d\tau_e \int \psi_{\text{vib}}' \mu_n \psi_{\text{vib}}'' \, d\tau_n \tag{10-10}$$

The second term vanishes since ψ_e' and ψ_e'' are orthonormal functions and for a given electronic transition $\int \psi_e' \mu_e \psi_e'' \, d\tau_e$ is a constant so that

$$\mu_{nm} \propto \int \psi_{\text{vib}}' \psi_{\text{vib}}'' \, d\tau_n$$

DISSOCIATION AND PREDISSOCIATION

Consider the following vibrational level change in an electronic transition as shown in Fig. 10-5. The vibrational energy of the molecule in the

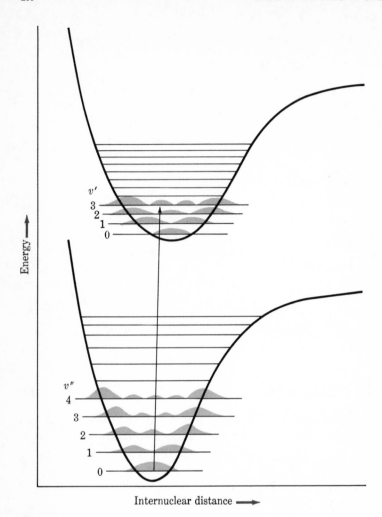

Fig. 10-4 Diagram showing the most probable transition. (*Adapted with permission from G. M. Barrow, "Introduction to Molecular Spectroscopy," McGraw-Hill Book Company, New York, 1962.*)

58 kcal

excited electronic state is so high that during the course of a vibration the internuclear distance approaches infinity, that is, the molecule dissociates. This corresponds to the *continuum* region of the spectrum. Figure 10-6 shows the electronic absorption spectrum of iodine vapor.[2] The vibrational bands converge at 4995 Å, which corresponds to a value of 5.7 kcal mole^{-1} for the dissociation energy. However, there is strong evidence to suggest that one of the iodine atoms formed is in the ground

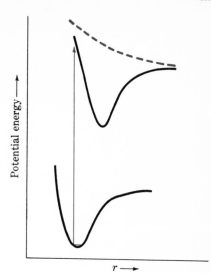

Fig. 10-5 Vibrational level change in an electronic transition which leads to dissociation.

state ($^2P_{\frac{3}{2}}$) and the other in the excited state ($^2P_{\frac{1}{2}}$). From the atomic spectra of iodine the separation between these two states is found to be 2.2 kcal g atom^{-1}. Thus the energy required for the following reaction

$$I_2 \xrightarrow{h\nu} 2I(^2P_{\frac{3}{2}})$$

is 3.5 kcal mole^{-1}. Similar studies have also been made on bromine, oxygen, and other molecules. It is important to note that this convergence-limit method is not generally applicable for obtaining dissociation energies since for many molecules the vibrational bands are too faint to be followed right up to the limit. In these cases the analytical extrapolation methods are often useful to obtain an estimate of the dissociation energies.[3]

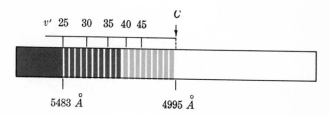

Fig. 10-6 Electronic absorption spectrum of iodine vapor. [*From "Molecular Spectra and Molecular Structure: I, Spectra of Diatomic Molecules," by Gerhard Herzberg, 2d ed. Copyright ©, by Litton Educational Publishing, Inc., by permission of Van Nostrand Reinhold Company.*]

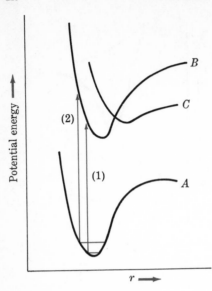

Fig. 10-7 Normal electronic transition (1) and electronic transitions that lead to predissociation (2).

In some molecules the potential energy curve of their excited electronic state has no minimum (see dotted curve in Fig. 10-5), and hence every transition will lead to dissociation. Consequently the electronic spectrum will simply be a continuum throughout.

Some electronic spectra show the unusual features that certain vibrational bands are well resolved but the rotational fine structures have a diffuse appearance. This is due to the predissociation phenomenon as follows: If there is more than one stable potential energy curve associated with the first excited electronic state, as shown in Fig. 10-7, transition 1 to state B will lead to the normal spectrum with vibrational as well as rotational fine structures. Transition 2, however, is quite different. Although the vibrational energy in this case is still not large enough to cause dissociation, the energy is large enough for the molecule to "crossover" to state C. There is no energy change accompanying this switching, and the process is known as *radiationless* transition. Since the time required for a complete vibration is about one-hundredth that for a complete rotation, the molecule will not have enough time to rotate before it dissociates in state C. Consequently only the vibrational fine structure can be observed. Predissociation spectra have been obtained for S_2, CaH, AlH, HgH and a few other molecules.[4]

SUGGESTIONS FOR FURTHER READING

Introductory

Dunford, H. B.: "Elements of Diatomic Molecular Spectra," Addison-Wesley Publishing Company, Inc., Reading, Mass., 1968.

Intermediate-advanced

3arrow, G. M.: "Introduction to Molecular Spectroscopy," McGraw-Hill Book Company, Inc., New York, 1962.

Herzberg, G.: "Molecular Spectra and Molecular Structure, I. Spectra of Diatomic Molecules," D. Van Nostrand Company, Inc., Princeton, N.J., 1950.

READING ASSIGNMENTS

3and Spectra and Dissociation Energies, F. E. Stafford, *J. Chem. Educ.*, **39**: 626 (1962).

Why Liquid Oxygen Is Blue, E. A. Ogryzlo, *J. Chem. Educ.*, **42**: 647 (1965).

REFERENCES

1. J. Franck, *Trans. Faraday Soc.*, **21**: 536 (1926). C. U. Condon, *Phys. Rev.*, **32**: 858 (1928).
2. Herzberg, p. 388.
3. A. G. Gaydon, "Dissociation Energies," p. 75, Chapman & Hall, Ltd., London, 1947.
4. Herzberg, p. 424.

11

Electronic Spectroscopy of Polyatomic Molecules

11-1 INTRODUCTION

The main difference between the electronic spectra of di- and polyatomic molecules lies in the resolution. In the latter case the rotational energy levels are more closely spaced because of the much larger moments of inertia; consequently, the rotational fine structure in general cannot be observed, and the vibrational structures are usually present only as broadbands. However, in spite of the poorer resolution, the electronic spectra of polyatomic molecules have provided very useful information of the electronic structure of a large number of compounds.

As is true with any other branch of spectroscopy, any satisfactory theory that deals with the electronic spectra must be able to account for the position and the intensity of the lines. In contrast to the molecular rotation and vibration problems, there is no adequate simple model. Most of the theoretical work, as we shall see in Sec. 11-3, is based on the molecular orbital theory.

Since the electronic transition may involve electrons in the σ, π, or

n orbitals,[†] it is possible, at least in principle, to have the following types of transitions: $\sigma \to \sigma^*$, $\pi \to \pi^*$, $n \to \pi^*$, and $n \to \sigma^*$. Transitions of the type $\pi \to \sigma^*$ and $\sigma \to \pi^*$ are forbidden by symmetry although they are known to take place in some compounds. Unfortunately there does not appear to be an accepted standard notation for describing the transitions. The following summarizes some of the notations currently favored by different groups of workers.[1]

Molecular orbital representation The transitions are denoted by $\pi^* \leftarrow \pi$, $\sigma^* \leftarrow \sigma$, $\pi^* \leftarrow n$, etc., and the corresponding excited states are referred to as (π^*,π), (σ^*,σ), (π^*,n). Both the arrows $\to$ and $\leftarrow$ are used to denote a transition, but this should not cause any confusion as long as the excited states are indicated.

Mulliken's symbols According to Mulliken the symbol N is used to denote the ground state and V, Q, R, etc., the excited states. $V \leftarrow N$ corresponds to the $\pi^* \leftarrow \pi$ transition, and $Q \leftarrow N$ to the $\pi^* \leftarrow n$ transition in the molecular orbital representations. $R \leftarrow N$ denotes the Rydberg transitions.[‡]

Platt's symbols In this classification the symbol A is used to denote the ground state, B the excited states involved in very high-intensity transitions, and L and C the partially forbidden transitions.

Group theory notations In this case a state is classified according to the behavior of the wave function (electronic) under the symmetry operations of the point group to which the molecule belongs. We shall discuss this in more detail in Sec. 11-3.

11-2 EXPERIMENTAL TECHNIQUES

The basic features of a UV spectrometer are:

1. A source of radiation. In the UV region (2000 to 3500 Å) this is usually a hydrogen discharge lamp; in the visible region (4000 to 6500 Å) this is a tungsten lamp.
2. A monochromator which is usually a prism or a grating.
3. A phototube or photomultiplier is the usual detector.
4. A pen recorder.

[†] n denotes the nonbonding orbital.

[‡] The Rydberg series here stemmed from atomic spectroscopy. In the molecular case the series arises when the electron is excited to the orbitals that are so far removed from the rest of the molecule that the singly charged molecular "core" plays the role of a nucleus.

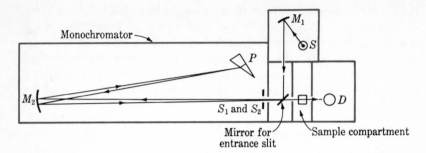

Fig. 11-1 Schematic diagram of a single-beam spectrometer. [*By permission of D. P. Shoemaker and C. W. Garland, "Experiments in Physical Chemistry," McGraw-Hill Book Company, New York, 1967.*]

In recording the spectra of transient species the photographic plate is often used. Figure 11-1 shows the schematic diagram of *single-beam* spectrometer.

11-3 THEORY

Let us consider first the electronic structure and spectrum of ethylene molecule—the simplest organic π system. The Hückel molecular orbital calculation gives the following bonding (ψ_1) and antibonding (ψ_2) molecular orbitals[2]

$$\psi_1 = \frac{1}{\sqrt{2}}(\phi_1 + \phi_2)$$

$$\psi_2 = \frac{1}{\sqrt{2}}(\phi_1 - \phi_2)$$

where ϕ_1 and ϕ_2 are the carbon $2p$ atomic orbitals. The ethylene molecule belongs to the D_{2h} point group—some of its symmetry elements are shown in Fig. 11-2. Figure 11-3 shows the bonding and antibonding

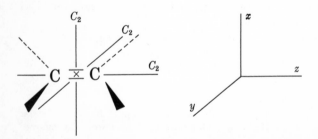

Fig. 11-2 Some of the symmetry elements of ethylene.

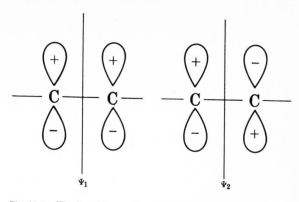

Fig. 11-3 The bonding and antibonding orbitals of ethylene.

orbitals of ethylene. If we now carry out the symmetry operations on ψ_1 and ψ_2 using the symmetry elements in the D_{2h} character table (Appendix 2), we obtain:

	E	$C_2{}^z$	$C_2{}^y$	$C_2{}^x$	i	σ^{xy}	σ^{xz}	σ^{yz}	
ψ_1	1	-1	-1	1	-1	1	1	-1	b_{3u}
ψ_2	1	-1	1	-1	1	-1	1	-1	b_{2g}

Hence the two molecular orbitals are classified as b_{3u} and b_{2g}. Next, we proceed to classify the electronic states as follows: In the ground state ψ_1 there are two electrons, and hence we have $(b_{3u})^2$ which belongs to the totally symmetric species of the D_{2h} group; that is, the electronic state is classified as ^1A_g. The superscript 1 means that the two electrons are paired in the ground state; that is, it is a singlet state. The first excited state is given by $b_{3u} \times b_{2g}$, which gives

E	$C_2{}^z$	$C_2{}^y$	$C_2{}^x$	i	σ^{xy}	σ^{xz}	σ^{yz}
1	1	-1	-1	-1	-1	1	1

which corresponds to B_{1u}. However, since the two electrons can either be parallel or antiparallel, we have the following two states: $^1B_{1u}$ and $^3B_{1u}$. In deriving these states we have adhered to the convention of using lowercase letters to indicate the symmetry species of the orbitals and capital letters to indicate the electronic states.

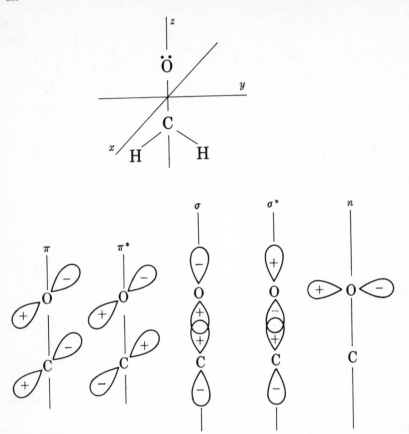

Fig. 11-4 The outer orbitals of formaldehyde.

As another example let us consider the electronic structure of formaldehyde (Fig. 11-4). The electronic structure for the ground state is

$$[(1s_C)^2(1s_O)^2(2s_O)^2(\sigma_{CH})^2(\sigma_{CH'})^2(\sigma_{CO})^2]\pi_{CO}{}^2 2p_O{}^2$$

The electrons in the brackets are of very low energy, and their excitations will not concern us here. Abbreviating π for π_{CO} and n for $2p_O$ and noting that formaldehyde belongs to the C_{2v} point group, we apply the symmetry operations for C_{2v} to the orbitals shown in Fig. 11-4. The results are

Orbital	Symmetry species
σ	a_1
σ^*	a_1
π	b_1
π^*	b_1
n	b_2

The electronic states of formaldehyde can then be evaluated in a similar manner to that shown for ethylene:

$(\pi)^2(n)^2$ $^1A_1$ ground state

$(\pi)^2(n)(\pi^*)$ $^1A_2$ or $^3A_2$ ⎫

$(\pi)(n)^2(\pi^*)$ $^1A_1$ or $^3A_1$ ⎬   excited states

$(\pi)^2(n)(\sigma^*)$ $^1B_2$ or $^3B_2$

$(\pi)(n)^2(\sigma^*)$ $^1B_1$ or $^3B_1$ ⎭

THE SELECTION RULES

Consider the following transitions for ethylene:

(a) $^3B_{1u}(\psi_1\psi_2) \leftarrow {}^1A_g(\psi_1)^2$

(b) $^1B_{1u}(\psi_1\psi_2) \leftarrow {}^1A_g(\psi_1)^2$

(c) $^1A_g(\psi_2)^2 \leftarrow {}^1A_g(\psi_1)^2$

The intensity of these transitions varies greatly due to the selection rule which can, in general, be written as $\int \psi_i \mathbf{M} \psi_f \, d\tau$, where ψ_i and ψ_f are the initial and final electronic states and $\mathbf{M}$ is the electronic-dipole-moment operator; that is, $\mathbf{M} = \Sigma e\mathbf{r}$, where $\mathbf{r}$ is the vector connecting the electron to some origin. The three components M_x, M_y, and M_z in the x, y, and z directions are related to $\mathbf{M}$ by the equation

$$M^2 = M_x^2 + M_y^2 + M_z^2$$

For a given point group, the assignments of M_x, M_y, and M_z to the appropriate symmetric species are given in the right-hand columns of the character table. They appear simply as x, y, and z. More specific selection rules are the following: (1) For molecules having a center of symmetry the selection rules are:

$g \leftarrow u$ $g \nleftarrow g$

$u \leftarrow g$ $u \nleftarrow u$

This would rule out transitions (a) and (b) for ethylene. (2) When there is a change in the multiplicity of the states, for example, singlet to triplet transitions, the spin wave functions must also be taken into account. When we consider the transition (a) for ethylene, we write the following integral I

$$I = \int \psi_1^2 \alpha\beta M \psi_1 \psi_2 \alpha\alpha \, d\tau_1 \, d\tau_2 \, d\sigma_1 \, d\sigma_2 \tag{11-1}$$

where α and β denote the two spin states and $d\sigma$ is the volume element for the spin part. Rearranging Eq. (11-1) we get

$$\overset{\text{Space}}{\int \psi_1^2 M \psi_1 \psi_2 \, d\tau_1 \, d\tau_2} \overset{\text{Spin}}{\int \alpha^3\beta \, d\sigma_1 \, d\sigma_2} = \int \psi_1^2 M \psi_1 \psi_1 \, d\tau_1 \, d\tau_2 \int \alpha\beta \, d\sigma_2 = 0$$

since

$$\int \alpha^2 \, d\sigma = 1 \qquad \text{and} \qquad \int \alpha\beta \, d\sigma = 0$$

Hence the selection rule is $\Delta S = 0$. Thus transition (a) for ethylene is very weak. For large molecules, especially those containing heavy atoms such as sulfur, iodine, etc., singlet to triplet transitions can give rise to quite strong lines because of the spin-orbit interaction.[3] (3) Lastly, we consider the symmetry of the space part of the wave function. For formaldehyde the following transitions are allowed from spin consideration alone:

(d) $^1A_2 \leftarrow {}^1A_1$
(e) $^1B_2 \leftarrow {}^1A_1$
(f) $^1B_1 \leftarrow {}^1A_1$
(g) $^1A_1 \leftarrow {}^1A_1$

We again examine integrals of the type†

$$\int \psi_i M \psi_f \, d\tau$$

From the symmetry of the ground and excited state it is clear that

$$^1A_1{}^1A_2 = {}^1A_2 \qquad {}^1A_1{}^1B_2 = {}^1B_2 \qquad {}^1A_1{}^1B_1 = {}^1B_1 \qquad {}^1A_1{}^1A_1 = {}^1A_1$$

Thus whether the product is totally symmetric or not depends on the symmetry of the upper state involved. Consider a specific transition, say, $^1A_2 \leftarrow {}^1A_1$. Since $^1A_2 \times {}^1A_1 = {}^1A_2$, and none of the components of M (M_x, M_y, or M_z) belongs to the species A_2 (Appendix 2), the integrand $\int \psi_i M \psi_f \, d\tau$ is an odd function and vanishes. This transition is forbidden

† The spin part of the wave function need not concern us here.

Table 11-1   Some of the allowed transitions of ethylene and formaldehyde in terms of the notations currently in use

	MO representations	Mulliken	Platt	Group theory
Ethylene	$\pi^* \leftarrow \pi$	$V \leftarrow N$	$B \leftarrow A$	$^1B_{1u} \leftarrow {}^1A_g$
Formaldehyde	$\pi^* \leftarrow \pi$	$V \leftarrow N$	$B \leftarrow A$	$^1A_1 \leftarrow {}^1A_1$
	$\sigma^* \leftarrow n$	$Q \leftarrow N$	$B \leftarrow A$	$^1B_2 \leftarrow {}^1A_1$

because of the symmetry of states. Similarly we can show that the following transitions are allowed by symmetry

$$^1A_1 \leftarrow {}^1A_1$$
$$^1B_2 \leftarrow {}^1A_1$$
$$^1B_1 \leftarrow {}^1A_1$$

These results are summarized in Table 11-1.

As a final example let us consider the electronic spectrum of the benzene molecule. This molecule has D_{6h} symmetry; the Hückel molecular orbital calculation gives the following results[4]

$$\psi_1 = \frac{1}{\sqrt{6}} (\phi_1 + \phi_2 + \phi_3 + \phi_4 + \phi_5 + \phi_6) \qquad a_{2u}$$

$$\psi_2 = \tfrac{1}{2}(\phi_2 + \phi_3 - \phi_5 - \phi_6) \qquad e_{1g}$$

$$\psi_3 = \frac{1}{\sqrt{12}} (2\phi_1 + \phi_2 - \phi_3 - 2\phi_4 - \phi_5 + \phi_6) \qquad e_{1g}$$

$$\psi_4 = \frac{1}{\sqrt{12}} (2\phi_1 - \phi_2 - \phi_3 + 2\phi_4 - \phi_5 - \phi_6) \qquad e_{2u}$$

$$\psi_5 = \tfrac{1}{2}(\phi_2 - \phi_3 + \phi_5 - \phi_6) \qquad e_{2u}$$

$$\psi_6 = \frac{1}{\sqrt{6}} (\phi_1 - \phi_2 + \phi_3 - \phi_4 + \phi_5 - \phi_6) \qquad b_{2g}$$

The dotted lines indicate the positions of the π-orbital nodes.   The energy level diagram is shown in Figure 11-5.

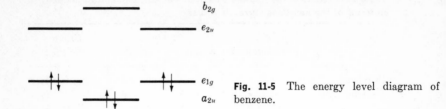

Fig. 11-5 The energy level diagram of benzene.

The ground electronic state is given by $(a_{2u})^2(e_{1g})^2(e_{1g})^2$, which belongs to the totally symmetric species $^1A_{1g}$. The lowest energy electronic transition gives the excited state $(a_{2u})^2(e_{1g})^2(e_{1g})(e_{2u})$, and from Fig. 11-5 it is clear that the following states should have the same energy.

$$
\left.\begin{array}{l}
\psi_1{}^2\psi_2{}^2\psi_3\psi_4 \\
\psi_1{}^2\psi_2{}^2\psi_3\psi_5 \\
\psi_1{}^2\psi_2\psi_3{}^2\psi_4 \\
\psi_1{}^2\psi_2\psi_3{}^2\psi_5
\end{array}\right\} \quad (a_{2u})^2(e_{1g})^2(e_{1g})(e_{2u})
$$

It is only necessary to consider the product $(e_{1g})(e_{2u})$ since $(a_{2u})^2(e_{1g})^2$ is totally symmetric. It can be shown by group theory that the product of $(e_{1g})(e_{2u})$ gives the symmetry species B_{1u}, B_{2u}, and E_{1u} for the excited states.[5] Note that E_{1u} denotes the doubly degenerate pair and that both the singlet and triplet states are possible in the excited state. Figure 11-6 shows the term level diagram for benzene. Figure 11-7 shows the electronic spectrum of benzene.

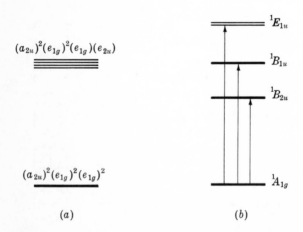

Fig. 11-6 The term level diagram of benzene. (a) No electron repulsion. (b) Electron repulsion taken into account.

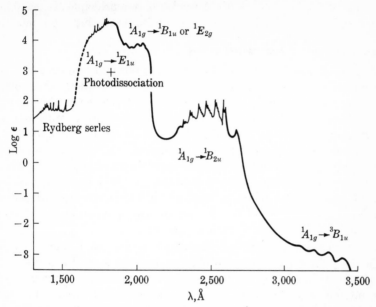

Fig. 11-7 Electronic spectrum of benzene. [*Kenneth S. Pitzer, "Quantum Chemistry,"* © *1953.   By permission of Prentice-Hall, Inc., Englewood Cliffs, New Jersey.*]

NTENSITY

For any branch of absorption spectroscopy the absorbance A is given by (Sec. 1-5)

$$A = \epsilon bc$$

In addition to the molar absorptivity ϵ quoted for an absorption band, however, a quantity called *oscillator strength* is also frequently employed. The integrated intensity I over a certain region in an electronic spectrum is given by

$$I = \int \epsilon \, d\nu \tag{11-2}$$

The oscillator strength f is defined as

$$f = \frac{2.3 \times 10^3 \, cm \int \epsilon \, d\nu}{Ne^2} = 4.33 \times 10^{-9} \int \epsilon \, d\nu \tag{11-3}$$

where the quantity $2.3 \times 10^3/Ne^2$ (N is Avogadro's number) is the calculated intensity for the $v = 0 \rightarrow 1$ transition for a harmonically bound electron.[6]   Allowed electronic transitions have f's of the order of 1 and

ϵ's about 10^4.   Of course the Franck-Condon principle discussed in Chap
10 applies also to polyatomic molecules in predicting the intensities.

POLARIZATION OF ABSORPTION BANDS

In the discussion of the selection rules for formaldehyde it was noted tha
some transitions are allowed only in the x direction, while others are
allowed in the y or z directions.   This selectivity cannot be distinguished
experimentally if ordinary light is used as the exciting source to randomly
oriented molecules.   The situation will be quite different if oriented
molecules (say in a crystal) are irradiated with linearly polarized light.
Consider the naphthalene molecule case shown in Fig. 11-8.   The lowest
energy electronic transition is polarized in the x direction which means
that $\int \psi_i M_x \psi_f \, d\tau \neq 0$, but $\int \psi_i M_y \psi_f \, d\tau$ and $\int \psi_i M_z \psi_f \, d\tau$ are both zero.[7]   The
next lowest transition is polarized in the y direction and hence only
$\int \psi_i M_y \psi_f \, d\tau$ is nonzero.   Thus if linearly polarized light is applied along
the x direction only the first transition would occur, and if it is applied
to the y direction, we would observe only the second transition.   If the
polarized light is applied between the x and y axes or if the molecules are
randomly oriented, then both transitions would occur.   It is obvious
that the polarization measurements are of considerable value in assigning
transitions.

FREE ELECTRON MODEL

For certain molecules such as conjugated dyes the simple free electron
(FE) model proposed by Kuhn[8a,b,c] is quite useful in analyzing their
electronic spectra.   Consider the dye molecule shown below

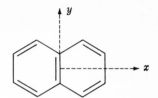

Fig. 11-8 The coordinate system for naphthalene.

We consider only the π electrons and assume that they are constrained to a one-dimensional potential well as shown in Fig. 11-9. L is the length of the box. The Schrödinger equation for the system is

$$-\frac{\hbar^2}{2m_e}\frac{d^2\psi}{dx^2} = E\psi \tag{11-4}$$

The wave functions are

$$\psi = \sqrt{\frac{2}{L}}\sin\frac{n\pi}{L}x \tag{11-5}$$

where $n = 1, 2, 3, \ldots$. The energies are given by

$$E_n = \frac{n^2h^2}{8m_eL^2} \tag{11-6}$$

The lowest energy transition for the dye molecule shown earlier is

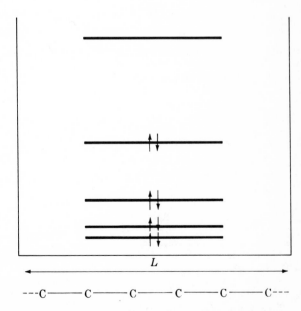

Fig. 11-9 The FE model for the conjugated dye system.

$n = 4 \to 5$; the energy change ΔE is given by

$$\Delta E = E_5 - E_4 = \frac{h^2}{8m_eL^2}(5^2 - 4^2) = \frac{9}{8}\frac{h^2}{m_eL^2}$$

In general, for a chain containing $2n$ atoms the energy of the lowest transition is given by

$$\Delta E = \frac{h^2}{8m_eL^2}[(n + 1)^2 - n^2] = \frac{h^2}{8m_eL^2}(2n + 1)$$

$$= \frac{h^2}{8m_eL^2}(N + 1) \tag{11-7}$$

where $N = 2n$. For the type of dye molecule shown earlier the two nitrogen atoms contribute three π electrons so that

$$N = p + 3$$

where p is the number of carbon π electrons. According to Kuhn L is the length of the chain (assumed to be straight), plus half a bond length (or one bond length if the chain is odd). Thus we have

Even chain: $L = Nr$
Odd chain: $L = (N + 1)r$

where r is the carbon-carbon distance. Equation (11-7), when expressed in wavelength λ, becomes

$$\lambda = \frac{8m_ecr^2}{h}\frac{(p + 3)^2}{p + 4} \tag{11-8}$$

Some of the results are summarized in Table 11-2.

Table 11-2 Absorption spectra of cyanine analogs

$Me_2N^+ = CH—(CH=CH)_r—NMe_2$

n	Wavelength (nm)	
---	Calculated	Observed
1	309	309
2	409	409
3	509	511

SOURCE: By permission from H. H. Jaffé and M. Orchin, "Theory and Applications of Ultraviolet Spectroscopy," John Wiley & Sons, Inc., New York, 1962.

Table 11-3   Spectra of α,ω-diphenylpolyenes
Ph(—CH=CH—)$_n$Ph

	λ_{max} (nm)			
n	Obs	$Calc$	$\epsilon_{max} \times 10^{-3}$	$Color$
1	306	310	24	colorless
2	334	334	40	colorless
3	358	358	75	pale yellow
4	384	380	86	greenish yellow
5	403	400	94	orange
6	420	420	113	brownish orange
7	435	438	135	copper bronze

SOURCE: Adapted, with permission, from H. H. Jaffé and M. Orchin, "Theory and Applications of Ultraviolet Spectroscopy," John Wiley & Sons, Inc., New York, 1962 and N. J. Juster, *J. Chem. Educ.*, **39**: 596 (1962).

For polyenes of the type H(—CH=CH)$_n$H, Eq. (11-7) does not apply, the reason being that the bond lengths in the two canonical struc-tures are no longer equal.   For example

$$CH_2=CH—CH=CH—CH=CH_2$$

$$\updownarrow$$

$$CH_2—CH=CH—CH—CH=CH_2$$

In this case the modified equation of Lewis and Calvin can be employed.[9] We write

$$(\lambda_{max})^2 = A + Bn \qquad (11\text{-}9)$$

where A and B are constants for a given series of polyenes and λ_{max} is the wavelength of maximum absorption.   A particularly interesting series is the α,ω-diphenylpolyenes, the results of which are summarized in Table 11-3.

CHROMOPHORES[10]

The term *chromophore* is usually taken to imply groups of atoms such as $>C=O$, $>C=C<$, $—N=N—$, and $—NO_2$ containing π electrons which are chiefly responsible for an absorption band.   The frequencies and the intensities of the bands are nevertheless sensitive to the nature of solvent and the inductive and resonance effects of the substituents. It is useful to define the following terms in regards to the change in frequency and intensity:

Bathochromic shift: Shift to lower frequencies (red shift)
Hyposochromic shift: Shift to higher frequencies (blue shift)
Hyperchromic effect: An increase in intensity
Hypochromic effect: A decrease in intensity

The term *auxochrome* applies to an atom or group of atoms which doe
not give rise to the absorption band on its own, but when conjugated to *
chromophore will cause a bathochromic shift and a hyperchromic effect
For example, the $>C=C<$ group in ethylene is the chromophore; whei
one of the hydrogen atoms is replaced by a halogen atom a bathochromi
shift and a hyperchromic effect are observed.   Thus the halogen atom
acts as the auxochrome.

It is interesting to examine the effect of chromophores on the
energies of transitions.   Hückel molecular orbital theory predicts tha
the transition frequency decreases with increase in conjugation.   This i
nicely shown by the α,ω-diphenylpolyene series (Table 11-3); as *
increases the polyenes become colored indicating that there is a batho
chromic shift (from UV to visible).

11-4 APPLICATIONS

Chemical analysis This discussion can be divided into two parts
qualitative and quantitative.   UV and visible spectra are used to identify
unknowns in a way analogous to NMR or IR.   However, because of the
much poorer resolution for the former, more care should be given in comparing the unknown with the standard sample.   In many cases the
chromophores can be identified by their characteristic frequencies.   As
mentioned earlier, the nature of the solvent, as well as temperature, can
significantly affect the frequencies and intensities.   For example, polar
solvents often stabilize the ground state of $\pi^* \leftarrow n$ transitions more than
the excited state, causing blue shifts:

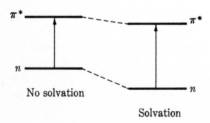

On the other hand, for $\pi^* \leftarrow \pi$ transitions the excited state is more stabilized and this leads to the red shifts:

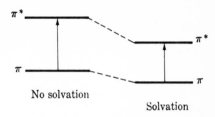

No solvation

Solvation

Since $\pi^* \leftarrow n$ transitions have a higher dipole change than $\pi^* \leftarrow \pi$ transitions, they are more solvent-dependent and shift to lower frequencies in
more polarizable solvents.

As with all other types of absorption spectroscopy, the quantitative
aspects of UV spectroscopy are based on the Beer-Lambert law. The
concentration of any compound with a known extinction coefficient can
be readily calculated from the measured absorbance A which is given by

$$A = \epsilon b c$$

Consider the following equilibrium between compounds X and Y

$$X \rightleftharpoons Y$$

If X and Y have absorption bands that overlap and if at some wavelength
over the overlapped region both X and Y have equal absorbance, then
there will be no change in the absorbance at this wavelength as the ratio
[Y]/[X] is varied. This gives rise to the *isosbestic point*. When there is
more than one isosbestic point present, we have

$$A_1 = c_X \epsilon_X^1 + c_Y \epsilon_Y^1$$
$$A_2 = c_X \epsilon_X^2 + c_Y \epsilon_Y^2$$

where 1 and 2 indicate the two different wavelengths. Since A_1 and A_2
can be measured, c_X and c_Y can be calculated (and hence the equilibrium
constant) if ϵ_X^1, ϵ_X^2, ϵ_Y^1, and ϵ_Y^2 are known. Pyridoxine, a molecule
related to vitamin B_6, undergoes the following reversible reaction at
pH < 7

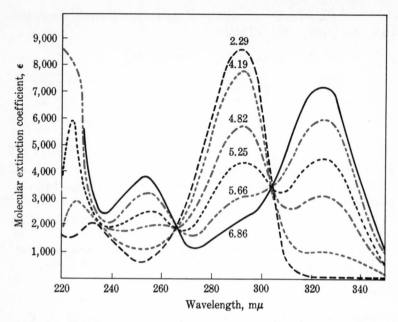

Fig. 11-10   The UV spectra of pyridoxine at various pH's.   [*From A. K. Lunn and R. A. Morton, The Analyst,* **77:** *718 (1952).   By permission of Dr. Morton and the Society for Analytical Chemistry.*]

Figure 11-10 shows the UV spectra at various pH's.   Note that there are three isosbestic points present.

As far as applications to biochemistry are concerned UV has a certain advantage over IR.[11]   Most biological systems have high water contents, and since water absorbs strongly in the IR its use is severely limited in some cases.   On the other hand aqueous solutions can be conveniently studied by UV.   Also for large biomolecules it is not easy or possible to obtain high resolution IR spectra, and hence it is difficult to follow small energy changes due to hydrogen bonding or other weak interactions.

Charge-transfer spectra   In 1948 Benesi and Hildebrand noticed that when iodine and benzene (and its methyl derivatives) were brought together in a 1:1 mole ratio, a new absorption band was observed (Fig. 11-11).[12]   According to Mulliken,[13] this new absorption band arises from the formation of a charge-transfer complex formed between a donor molecule D (benzene) and an acceptor molecule A (iodine).   The com-

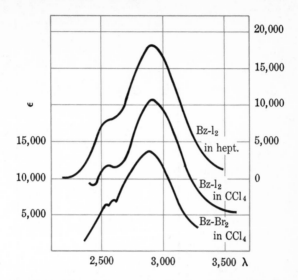

Fig. 11-11 The charge-transfer bands of halogens with benzene. [*From N. S. Ham, J. R. Platt, and H. M. McConnell, J. Chem. Phys.,* **19:** *1301 (1951). By permission of Dr. McConnell and Dr. Platt and the American Institute of Physics.*]

plex can be described by the resonance structures

$$DA \leftrightarrow D^+A^-$$
covalent   ionic

The total wave function of the ground state ψ_0 is

$$\psi_0 = \psi(DA) + \lambda\psi(D^+A^-)$$

where λ^2 is a measure of the amount of charge transfer. The benzene-iodine complex cannot be isolated as a solid and is partially dissociated in solution. The charge-transfer band arises from the electronic transition from the ground state ψ_0 to the first excited state ψ_1, which is given by

$$\psi_1 = \psi(DA) + \lambda'\psi(D^+A^-)$$

where $\lambda' > \lambda$, that is, the excited state is more ionic.

The tendency of charge-transfer-complex formation and the energy change in the transition must depend on the ionization potential of D and the electron affinity of A, and such correlations have indeed been made.[14]

The quantity of interest here is of course the equilibrium constant

Table 11-4 Values of λ_{max} and ϵ_c for charge-transfer absorption of alkylbenzene complexes[†]

Acceptors → donor (I_p, eV)[‡]	s-$C_6H_3(NO_2)_3$			$2(NC)C=C(CN)_2$			I_2			ICl		
	K	λ_{max} $(m\mu)$	ϵ_{DA}	K	λ_{max} $(m\mu)$	ϵ_{DA}	K	λ_{max} $(m\mu)$	ϵ_{DA}	K	λ_{max} $(m\mu)$	ϵ_{DA}
Benzene (9.24)	0.82	284	9755	3.0	384	3570	0.15	292	16400	0.54	282	8130
Toluene (8.82)	1.82	306	4350	3.7	406	3330	0.16	302	16700	0.87	288	8000
o-Xylene (8.58)	2.08	312	4080	7.0	430	3860	0.27	316	12500	1.24	298	7870
m-Xylene (8.6)	....	....	....	6.0	440	3300	0.31	318	12500	1.39	298	9180
p-Xylene (8.48)	....	....	....	7.6	460	2650	0.31	304	10100	1.51	292	6540
Mesitylene (8.14)	2.67	335	3270	17.3	461	3120	0.82	332	8850	4.59	307	7870

[†] In all cases the units for ϵ_{DA} are cm⁻¹ mole⁻¹l. For the s-$C_6H_3(NO_2)_3$ complexes the K values reported are in (mole fraction)⁻¹ l.
[‡] The figures in parentheses are the ionization of potentials of the donors.
SOURCE: By permission of L. J. Andrews and R. M. Keefer, "Molecular Complexes in Organic Chemistry," Holden-Day, Inc., Publisher, San Francisco, 1964.

K for the reaction†

$$D + A \rightleftharpoons DA$$

Thus

$$K = \frac{[DA]}{[D][A]}$$

The general procedure is to measure the absorbance d of the complex as a function of the concentration of the donor, which is present in large excess of the acceptor. We have

$$d = \epsilon_{DA}b[DA]$$

$[D]_0$ and $[A]_0$ are the initial concentrations and since $[D]_0 \gg [A]_0$, we have

$$K = \frac{[DA]}{[D]_0\{[A]_0 - [DA]\}}$$

Equating $[DA]$ for the last two equations and rearranging, we get

$$\frac{1}{d} = \frac{1}{\epsilon_{DA}[A]_0 b} + \frac{1}{K\epsilon_{DA}[A]_0 b}\frac{1}{[D]_0}$$

Thus a plot of $1/d$ versus $1/[D]_0$ gives a straight line and from the slope and intercept both ϵ_{DA} and K can be obtained. Table 11-4 gives the λ_{max} and K's for some alkylbenzene charge-transfer complexes.

Chemical kinetics UV spectroscopy can be conveniently employed in kinetic studies. Thus by monitoring the intensity of a certain absorption peak with time the rate of a chemical reaction can be readily followed. The restrictions are that the rates must be relatively slow (half-lives of the order of a minute) and that the spectra of reactants and products do not overlap completely. For fast reactions the so-called stopped-flow

† We now write DA to denote the charge-transfer complex.

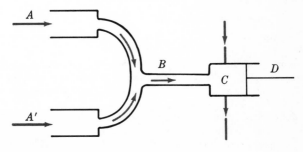

Fig. 11-12 Rapid stop-flow apparatus for kinetic studies.

method can be employed.[15a,b]   In this arrangement two solutions (reactant and product) are passed through the reaction chamber B and the flow of the mixed solution is stopped by the piston D (Fig. 11-12).   The absorbance of either species is monitored photometrically at C.   The detector is usually a photomultiplier whose output is displayed on the oscilloscope screen with a time base.   With suitable adjustment of flow rates, reactions with half-lives down to milliseconds can be studied.

PROBLEMS

11-1. The electronic spectra of ethylene and alkylated ethylenes show a strong absorption band at 7 eV which is assigned as a $\pi^* \leftarrow \pi$ transition.   There is also a much weaker band found in the 5-to 6-eV region.   Convert the electronvolts to the angstrom unit and make a reasonable assignment for the weak transition.

11-2. Many aromatic hydrocarbons are colorless, but their cation and anion radicals are often strongly colored.   For example, naphthalene is colorless but its anion radical is green.   Give a qualitative explanation.   Also, would you expect the anion and cation radical of naphthalene to have similar color?   Explain.

11-3. The λ_{max} of a compound decreases when the solvent changes from water to ethanol to hexane.   Deduce the nature of the levels involved in the transition.

11-4. Many amines react with quinhydrone to form deeply colored solutions.   Suggest an experiment to test whether the color is due to the charge-transfer complex or normal reaction products.

11-5. Although the UV technique can be conveniently employed in kinetic studies in general, it is not suited to study the rate of many intra- and intermolecular processes, in contrast to NMR.   Explain.

11-6. The biphenyl molecule absorbs at 246 nm with a molar absorptivity of 16,300; the 2,4,6,2',4',6'-hexamethylbiphenyl molecule absorbs at 267 nm with a molar absorptivity of 545.   Also, mesitylene is known to absorb at 266 nm with a molar absorptivity of 260.   Comment on the structure of the hexamethylbiphenyl and account for the shift in wavelength.

11-7. The UV spectrum of a mixture of *cis*- and *trans*-stilbene shows two peaks at 278 and 294 nm, respectively.   On what basis would you make the assignment?

SUGGESTIONS FOR FURTHER READING

Intermediate-advanced

Barrow, G. M.: "Introduction to Molecular Spectroscopy," McGraw-Hill Book Company, New York, 1962.

Bauman, R. P.: "Absorption Spectroscopy," John Wiley & Sons, Inc., New York, 1962.

Herzberg, G.: "Molecular Spectra and Molecular Structure.   III. Electronic Spectra and Electronic Structure of Polyatomic Molecules," D. Van Nostrand Company, Inc., Princeton, N.J., 1966.

Jaffé, H. H., and M. Orchin: "Theory and Applications of Ultraviolet Spectroscopy," John Wiley & Sons, Inc., New York, 1962.

Sandorfy, C.: "Electronic Spectra and Quantum Chemistry," Prentice-Hall, Inc., Englewood Cliffs, N.J., 1964.

Suzuki, H.: "Electronic Absorption Spectra and Geometry of Organic Molecules," Academic Press, Inc., New York, 1967.

READING ASSIGNMENTS

A Molecular Spectral Corroboration of Elementary Operator Quantum Mechanics, R. E. Gerkin, *J. Chem. Educ.*, **42**: 490 (1965).
Applications of Absorption Spectroscopy in Biochemistry, G. R. Penzer, *J. Chem. Educ.*, **45**: 692 (1968).
The Fates of Electronic Excitation Energy, H. H. Jaffé and A. L. Miller, *J. Chem. Educ.*, **43**: 469 (1966).
Understanding Ultraviolet Spectra of Organic Molecules, H. H. Jaffé, D. L. Beveridge, and M. Orchin, *J. Chem. Educ.*, **44**: 383 (1967).
Color and Chemical Constitution, N. J. Juster, *J. Chem. Educ.*, **39**: 597 (1962).

REFERENCES

1. J. N. Pitts, Jr., F. W. Wilkinson, and G. S. Hammond, The Vocabulary of Photochemistry, in W. A. Noyes, Jr., G. S. Hammond, and J. N. Pitts, Jr. (eds.), "Advances in Photochemistry," vol. 1, Interscience Publishers, a division of John Wiley & Sons, Inc., New York, 1963.
2. A. Streitwieser, Jr., "Molecular Orbital Theory for Organic Chemists," p. 39, John Wiley & Sons, Inc., New York, 1961.
3. Jaffé and Orchin, p. 145.
4. F. A. Cotton, "Chemical Applications of Group Theory," p. 125, Interscience Publishers, a division of John Wiley & Sons, Inc., New York, 1963.
5. Barrow, p. 285.
6. Barrow, p. 81.
7. Sandorfy, p. 342.
8a. H. Kuhn, *J. Chem. Phys.*, **17**: 1198 (1949).
8b. N. S. Bayliss, *Quart. Rev.*, **6**, 319 (1952).
8c. Jaffé and Orchin, p. 220.
9. G. N. Lewis and M. Calvin, *Chem. Rev.*, **25**: 273 (1939).
10. Jaffé and Orchin, chap. 9.
11. R. A. Morton, Spectrophotometry in the Ultraviolet and Visible Regions, in M. Florkin and E. H. Stotz (eds.), "Comprehensive Biochemistry," vol. 3, Elsevier Publishing Company, Amsterdam, 1962.
12. H. A. Benesi and J. H. Hildebrand, *J. Am. Chem. Soc.*, **70**: 2832 (1948).
13. R. S. Mulliken, *J. Am. Chem. Soc.*, **72**: 600 (1950), **74**: 811 (1952); *J. Phys. Chem.*, **56**: 801 (1952).
14a. S. H. Hastings, J. L. Franklin, J. C. Schiller, and F. A. Matsen, *J. Am. Chem. Soc.*, **75**: 2900 (1953).
14b. Ref. 2 (Streitwieser), p. 199.
15a. E. F. Caldin, "Fast Reactions in Solution," p. 44, John Wiley & Sons, Inc., New York, 1964.
15b. B. Chance, R. H. Eisenhardt, Q. H. Gibson, and K. K. Lonborg-Holm (eds.), "Rapid Mixing and Sampling Techniques in Biochemistry," p. 89, Academic Press, Inc., New York, 1964.

12

Photoluminescence: Fluorescence and Phosphorescence

12-1 GENERAL DISCUSSION

The word *photoluminescence* means the use of photons to excite light emission and it therefore includes Raman and Rayleigh scattering as well as fluorescence and phosphorescence. Here we shall be concerned only with the latter two phenomena.

If the excitation of a molecule by light does not lead to chemical reaction or energy transfer by collision, then the molecule would eventually return to the ground state with the release of energy; that is, this process shows up as luminescent emission. There are two major paths for the excited molecule to reach the ground state; they give rise to either the fluorescence or phosphorescence phenomenon. The former is a short-lived emission, lasting from 10^{-8} to 10^{-5} sec while the latter is a long-lived emission, lasting from 10^{-3} to the order of seconds.

FLUORESCENCE

When molecules are radiated with light of the appropriate frequencies they can undergo absorption from the ground to the first excited electronic state. Although at room temperature most molecules are in their ground vibrational level, after the excitation they can end up in any one of the vibrational levels in the first excited electronic state. Since all these excited states are unstable the molecules can *directly* return to the ground electronic state with the emission of light and hence fluorescence spectra are observed. Thus we can think of fluorescence as being the reverse process of absorption. However, a comparison of the absorption and fluorescence spectrum of the same compound shows that they do not superimpose on each other as expected; rather, they are the mirror images of each other, with the emission spectrum displaced towards the longer wavelength. The reason is that since the time required to execute a vibration (about 10^{-13} sec) is much shorter than the decay or mean lifetime (about 10^{-9} sec), most of the excess vibrational energy will be dissipated to the surroundings, and the excited molecules will now decay in their *ground* vibrational levels. The situation is illustrated in Fig. 12-1.

We define fluorescence yield ϕ as the ratio of the photons emitted by fluorescence to the total photons absorbed. The intensity of emitted radiation after the exciting light is turned off for time t is given by

$$I(t) = I_0 e^{-t/\tau} \tag{12-1}$$

where τ is the average decay period. Note that when $t = \tau$, $I(t) = I_0/e$.

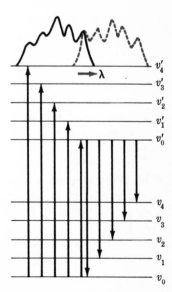

Fig. 12-1 The relation between absorption and fluorescence. [*By permission from C. Reid, "Excited States in Chemistry and Biology," Academic Press, Inc., New York, 1957.*]

I_0 is the intensity of emitted light at $t = 0$. The decay time is given by

$$\tau = \frac{1.5}{f\nu^2} \qquad \text{sec} \tag{12-2}$$

where f is the oscillator strength for the absorption.

If fluorescence is not the only available path for decay, as is usually the case, ϕ is less than one and we say that the fluorescence is quenched. Some of the quenching mechanisms are:

1. *Internal conversion* For example, absorption of light can lead to pre-dissociation if the bond dissociation energy is less than that for the transition. For this reason fluorescence normally does not occur in the vacuum UV region since the energy involved is too large. This therefore excludes most saturated compounds which undergo $\sigma^* \leftarrow \sigma$ transitions.

2. *External conversion* Collisions between solute-solvent and solute-solute are alternative ways for deactivation. In solutions such quenching must obviously depend on the viscosity of the solution and this is indeed the case. For example, for most compounds ϕ's are found to be larger in glycerol than ethanol. The concentration of the fluorescent solute can also provide a mechanism for quenching known as *concentration quenching*, as is the situation when the solute molecules form nonfluorescent dimers or higher aggregates and ϕ decreases with the increase in concentration.

3. *Excitation transfer* This is a quantum mechanical effect which involves the transfer of energy without radiation or collision. It depends on the overlap of the emission band of the excited molecule and the absorption band of the fluorescence-quenching molecule. When the quenching molecule dissipates the energy as light we call it sensitized fluorescence. An example of this was provided by Cario and Franck who irradiated a mixture of mercury and thallium vapor with the mercury line at 2537 Å.[1] Although thallium is transparent to this radiation the sensitized fluorescence from the excited states of thallium was observed.

PHOSPHORESCENCE

Phosphorescence offers a different path for the excited molecule to return to the ground state with the emission of light. In general it is easily distinguished from fluorescence because: (1) It has a much longer decay period. (2) Its spectrum is not the mirror image of the absorption spectrum. (3) It is not observable in solutions at room temperature. (4) It is rarely observed in gases. The mechanism of phosphorescence decay was first given by Lewis and Kasha in 1944 who showed that it involves

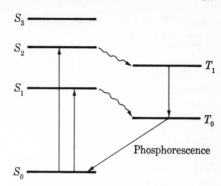

Fig. 12-2 The triplet decay scheme.

the molecule decaying via a metastable state, which is a triplet.[2] The decay scheme is shown in Fig. 12-2.

The decay from the higher singlets to the triplets is by the radiationless transition, which probably involves the crossing of the potential energy curves of the type discussed in Chap. 10 (see *predissociation*). The decay from the triplet to the ground state singlet is forbidden by spin symmetry and is therefore slow. However, the discussion of multiplicity is meaningful only in the Russell-Saunders coupling scheme. In practice there is always some spin-orbit interaction to "mix" the states so that triplet ⇌ singlet transitions are allowed although we expect them to be weaker than singlet-singlet transitions. According to Hund's rule the triplet level always lies lower then the corresponding singlet level and for this reason phosphorescence spectra occur at longer wavelengths compared with the absorption and fluorescence spectra. The most direct evidence of singlet ← triplet decay mechanism is provided by the magnetic susceptibility[3] and ESR measurements (see Chap. 6). These experiments show that by building up a steady concentration with continuous irradiation the phosphorescent molecules are paramagnetic and the ESR spectra can be analyzed only if we assume them to be in the triplet state. Table 12-1 shows the phosphorescence properties of some molecules. Note that the phosphorescence decay periods of halogenated hydrocarbons are much shorter than the hydrocarbons because of the enhanced spin-orbit interaction (presence of heavy atoms) which leads to greater singlet ← triplet transition probabilities.

There are also a number of mechanisms which can deactivate the triplet states. They are:

1. *Collision* In solution the time required for molecular collision is short compared with the triplet mean lifetime and consequently triplet molecules will lose their energy by colliison and not by radiation.

**Table 12-1 Phosphorescence lifetime
of some compounds**

Compound	τ (sec)
Benzene†	8
Toluene	8.8
Naphthalene	2.5
d-Naphthalene	19
Fluoronaphthalene	1.5
Chloronaphthalene	0.3
Bromonaphthalene	0.02
Iodonaphthalene	0.003

† The decay of benzene is quite complex and may not be exponential. The
value given here should only be taken
as an order of magnitude.

2. *Paramagnetic ion quenching*   Phosphorescence is quenched in the
   presence of paramagnetic ions. It is believed that they favor the
   radiationless transition of the triplet to the ground state singlet
   through the formation of a complex.
3. *Spin inversion*   The transfer of the electronic energy of a triplet molecule ^3X to a singlet molecule ^1Y can take place as follows:

$$^3X + {}^1Y \rightarrow {}^1X + {}^3Y$$

   Such a process is allowed by the spin-conservation rules and can
   take place as long as the triplet state of Y lies lower than that of X.
   An example of this is the triplet transfer from benzophenone to
   naphthalene.[4]

These three effects can be eliminated to a large extent in rigid solutions.

SINGLET–TRIPLET AND TRIPLET–TRIPLET TRANSITIONS

Just as the singlet-to-singlet absorption is the reverse of fluorescence, we
also have singlet-to-triplet absorption, which is the reverse of phosphorescence. Such transitions, as expected, give rise to very weak lines.
The first such transition was observed by Sklar in benzene.[5] It is interesting to note that the same type of mirror images exist in general between
singlet-triplet and phosphorescence spectra.

Analogous to the singlet states we can also label the triplet states
as *ground, first-excited, second-excited state*, etc. In phosphorescence the
decay is always from the ground triplet state to the ground singlet state.
However, by maintaining a steady concentration of the triplet state it is

possible to induce transition from this ground triplet state to the higher ones. The first such observation was made by Lewis and his coworkers on fluorescein.[6] They found two absorption bands corresponding to $T_1 \leftarrow T_0$ and $T_2 \leftarrow T_0$ transitions. Vibrational analyses of the triplet-triplet absorptions have yielded much useful information regarding the geometry of the triplet molecules and their interaction with the environment.

12-2 EXPERIMENTAL TECHNIQUES

FLUORESCENCE

The sample studied is usually in solution, and measurements are carried out at room temperature. The setup is quite analogous to that for Raman since both measure scattered radiations. The source, which does not have to be monochromatic, is either a tungsten or mercury lamp. The detector can either be a photomultiplier or a photographic plate and is placed at right angles to the incident radiation. An alternative way to observe fluorescence makes use of the *excitation spectrum*. In general the fluorescence intensity is proportional to the intensity of the exciting light, the extinction coefficient ϵ of the corresponding absorption, and ϕ. If we keep the intensity of the exciting light constant and vary the frequency,† the fluorescence intensity will then be proportional to ϵ. Thus the excitation spectrum we obtain corresponds closely to the absorption spectrum. This setup differs from the fluorescence experiments in that a monochromator is now placed between the source and the sample. The excitation spectrum has the following advantages over fluorescence: (1) It is more sensitive—concentrations as low as 10^{-8} M can be detected. (2) In a mixture of absorbing compounds the absorption spectrum of one particular compound can be singled out by monitoring to the appropriate fluorescence band.

PHOSPHORESCENCE

Figure 12-3 shows the essential features for observing phosphorescence. The sample is alternately illuminated by the source and analyzed by the use of a chopper (not shown). In this manner, the phosphorescence emission can be studied relatively free of interference from fluorescence and scattered lights. No phosphorescence can be observed in solution at room temperature since the high collision rate between molecules deactivates the photoexcited state. Usually spectra are recorded with compounds maintained at liquid nitrogen or lower temperatures; thus it is imperative to use solvents which when frozen form a clear, transparent glass suitable for optical studies. Examples of these are 2-methyltetra-

† ϕ is roughly independent of the exciting frequency.

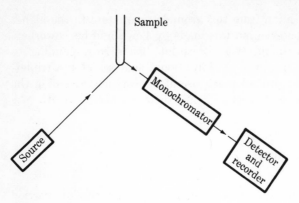

Fig. 12-3 Schematic diagram for observing phosphorescence.

hydrofuran or EPA, which is a mixture of ether, isopentane, and ethyl alcohol in the ratio of 5:5:2 by volume.

12-3 APPLICATIONS

In principle both fluorescence and phosphorescence techniques can be used for compound identifications. However, at the present time the latter is mainly used to study the electronic structure of the triplet state rather than as an analytical tool. On the other hand, because of its high sensitivity, the fluorescence technique has great practical applications.

The fluorescence technique is now being employed in increasing amounts in biochemistry. For example, flavin mononucleotide (FMN) is strongly fluorescent in its oxidized form. Its fluorescence is quenched, sometimes completely, when it is bound to apoenzymes.[7] This therefore affords a convenient method for measuring the equilibrium. The high sensitivity of the technique also makes the kinetic studies of a large number of systems possible since the rate of a biochemical reaction can be sufficiently slowed down for measurement by employing very dilute solutions.

Useful information regarding the conformation of proteins and other macromolecules can sometimes be obtained by using the fluorescence probes.[8] These probes are compounds whose fluorescent properties (quantum yield, lifetime, etc.) are affected by their immediate environments. A typical example is the 2-p-toluidinylnaphthalene-6-sulfonate (2,6-TNS), which is called a hydrophobic probe because its quantum yield is much higher in nonpolar solvents than polar solvents. In its interaction with proteins 2,6-TNS is preferentially bound to the hydrophobic chains. For example, the quantum yield of 2,6-TNS in native bovine serum albumin (BSA) is much lower than that when the protein is denatured. The results suggest that in the native state the hydrophobic chains of BSA are more exposed to water, whereas in the denatured state, the chains are more folded up to exclude water. Note that the principle of using fluorescence probes to study conformation is quite analogous to the spin-label technique discussed in Sec. 6-4.

PROBLEMS

12-1. Explain why fluorescence is in general a more sensitive technique than absorption.

12-2. How would you find out whether certain bands observed in a phosphorescence spectrum are due to impurity?

12-3. Why is the shape of a fluorescence spectrum independent of the wavelength of the exciting source.

12-4. Verify Eq. (12-3) (use Einstein probability of spontaneous emission and the definition of oscillator strength).

12-5. Design an experiment that would allow you to measure the quantum yield of phosphorescence using the ESR technique.

12-6. The phosphorescence decay of a certain compound is recorded photographically; the following data are obtained:

Intensity	1	0.819	0.670	0.540	0.449	0.301	0.165
Time (sec)	0	2	4	6	8	12	18

What is the lifetime of the triplet state?

SUGGESTIONS FOR FURTHER READING

Intermediate-advanced

Ehrenberg, A., and H. Theorell in M. Florkin and E. H. Stotz (eds.): "Comprehensive Biochemistry," vol. 3, Elsevier Publishing Company, Amsterdam, 1962.
El-Sayed, M. A.: *Acc. Chem. Res.*, **1:** 8 (1968).
Jaffé, H. H., and Orchin, M.: "Theory and Applications of Ultraviolet Spectroscopy," chap. 19, John Wiley & Sons, Inc., New York, 1962.
Lower, S. K., and M. A. El-Sayed: *Chem. Rev.*, **66:** 199 (1966).
Reid, C.: "Excited States in Chemistry and Biology," Academic Press, Inc., New York, 1957.
West, W.: "Chemical Applications of Spectroscopy," chap. 6, Interscience Publishers, Inc., New York, 1956.

READING ASSIGNMENTS

A Molecular Fluorescence Experiment for Undergraduate Physical Chemistry, J. I. Steinfeld, *J. Chem. Educ.*, **42:** 85 (1965).

Some Aspects of Fluorescence and Phosphorescence Analysis, D. M. Hercules, *Anal. Chem.*, **38**: 29A (1966).

The Selection of Optimum Conditions for Spectrochemical Methods. II, W. J. McCarthy and J. D. Winefordner, *J. Chem. Educ.*, **44**: 137 (1967).

The Selection of Optimum Conditions for Spectrochemical Methods. IV, J. J. Cetorelli, W. J. McCarthy, and J. D. Winefordner, *J. Chem. Educ.*, **45**: 99 (1968).

Two Fluorescence Experiments, S. F. Russo, *J. Chem. Educ.*, **46**: 375 (1969).

The Triplet State, N. J. Turro, *J. Chem. Educ.*, **46**: 2 (1969).

Liquid Scintillation Counting, W. Yang and E. K. C. Lee, *J. Chem. Educ.*, **46**: 277 (1969).

Interaction of a Naphthalene Dye with Apohemoglobin, Fancolli, T. M., and S. F. Russo, *J. Chem. Educ.*, **47**: 54 (1970).

REFERENCES CITED

1. G. Cario and J. Franck, *Z. Physik*, **17**: 202 (1923).
2. G. N. Lewis and M. Kasha, *J. Am. Chem. Soc.*, **64**: 1916 (1942).
3. D. F. Evans, *Nature*, **176**: 777 (1955).
4. A. Terenin and V. Ermolaev, *Trans. Faraday Soc.*, **52**: 1042 (1956).
5. A. L. Sklar, *J. Chem. Phys.*, **5**: 669 (1937).
6. G. N. Lewis, D. L. Lipkin, and T. T. Magel, *J. Am. Chem. Soc.*, **63**: 3005 (1941).
7. S. Shifrin and N. O. Kaplan, in F. F. Nord (ed.), "Advances in Enzymology," vol. 22, Interscience Publishers, Inc., New York, 1960.
8. G. M. Edelman and W. O. McClure, *Acc. Chem. Res.*, **1**: 65 (1968).

13

Optical Rotatory Dispersion and Circular Dichroism

13-1 THEORY

POLARIZED LIGHT

We saw in Chap. 1 that light is just one form of electromagnetic radiation which has an electric component and a mutually perpendicular magnetic component. Both components oscillate in space and the planes containing these components are perpendicular to the direction of propagation. In the case of natural or unpolarized light the instantaneous polarization, that is, the direction of the electric and magnetic components in space, changes rapidly in a random manner. However, if this natural light is passed through a sheet of Polaroid or a Nicol prism, the outcoming light would have its electric and magnetic components confined to certain particular directions, as shown in Fig. 13-1. This is called *linearly polarized*, or *plane-polarized, light*.

The electric field vector E can be thought of as being made up of two component vectors E_L and E_R, which correspond to the left and right *circularly* polarized waves. Figure 13-2a shows the variation of the

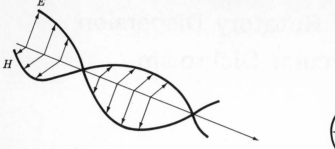

Fig. 13-1   Linear- or plane-polarized light.

electric field vector E with time and Fig. 13-2b shows the variation of E as the resultant of the two rotating vectors.

In an ordinary medium vectors E_L and E_R rotate at the same speed and hence E is confined to the x plane.   In an optically active medium E_L and E_R rotate at different speeds and although the light is still plane-polarized, the plane containing E now makes an angle α with the x axis (Fig. 13-3).   This rotation of the plane of polarization arises from the fact that in an optically active medium the refractive index n is different for the left and right circularly polarized light; that is, $n_L \neq n_R$, and hence the different speeds of rotation.   The medium is said to be *circularly birefringent,* and the angle of rotation per unit length is given by[1]

$$\alpha = \frac{1800}{\lambda} (n_L - n_R) \tag{13-1}$$

where λ is the wavelength of the incident light.   In practice we measure either the *specific rotation* $[\alpha]$, which is defined as

$$[\alpha] = \frac{\alpha \times 100}{l \times c} \tag{13-2}$$

where α is in degrees, l is in decimeters, and c is in g 100 ml^{-1} of solution, or we measure the *molecular rotation* $[\Phi]$ which is defined as

$$[\Phi] = \frac{[\alpha]M}{100} \tag{13-3}$$

where M is the molecular weight of the optically active compound. Sometimes $[\alpha]$ is expressed as $[\alpha]_\lambda{}^T$ where T is the temperature.   If the sense of rotation of the plane of polarization is to the right, the substance (medium) is called dextrorotatory, denoted by $(+)$; if to the left, levo-rotatory, denoted by $(-)$.

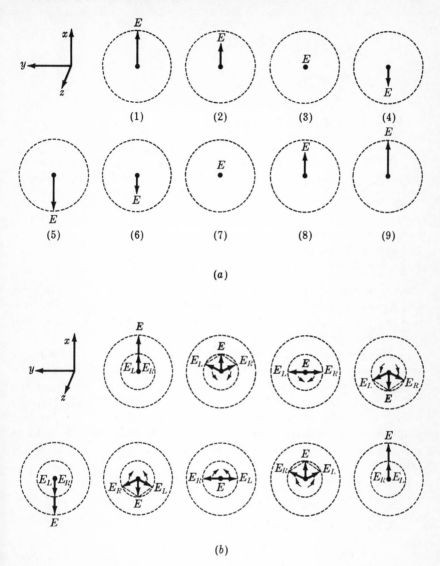

Fig. 13-2 (*a*) Variation of *E* with time. (*b*) Variation of *E* as a resultant of two rotating vectors E_L and E_R. [*By permission of C. Djerassi, "Optical Rotatory Dispersion," McGraw-Hill Book Company, New York, 1960.*]

OPTICALLY ACTIVE MOLECULES

Before we discuss the rotation of plane-polarized light by optically active molecules, it is useful to first define the criteria for optical activity. To be optically active, a molecule must *not* possess any one of the following symmetry elements:

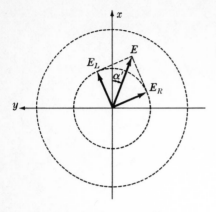

Fig. 13-3 Rotation of E in an optically active medium. [*By permission of C. D. Djerassi, "Optical Rotatory Dispersion," McGraw-Hill Book Company, New York, 1960.*]

1. Center of symmetry (i)
2. Plane of symmetry (σ)
3. An improper axis (S)

Since a onefold alternating axis is identical with a plane of symmetry and a twofold alternating axis is identical with a center of inversion, that is, $\sigma \equiv S_1$, and $i \equiv S_2$, (1), (2), and (3) are reduced to the statement that the molecule must not have an improper axis.

It is also important to distinguish the terms *symmetric, dissymmetric,* and *asymmetric* in discussing optical activity. A molecule is said to be symmetric (and hence optically inactive) if it possesses an improper axis S_n. A molecule is said to be dissymmetric (and hence usually optically active) if it lacks an improper axis S_n and may or may not contain a simple axis of symmetry, for example, C_2. A molecule is said to be asymmetric (and hence optically active) if it lacks both an improper axis and a simple axis.

Some examples of optically active molecules are

$$
\begin{array}{cc}
\text{I} & \text{CH}_3 \\
| & | \\
\text{CH}_3\!-\!\text{C}\!-\!\text{Cl} & \text{H}\!-\!\text{C}\!-\!\text{C}_3\text{H}_7 \\
| & | \\
\text{Br} & \text{C}_2\text{H}_5 \\
(a) & (b)
\end{array}
$$

The refractive index is related to the polarizability of the molecule; it seems logical, therefore, that optical activity should be related to the same quantity. Brewster found that the center of optical activity can be described as an asymmetric screw pattern of polarizability.[2] In (a) all four atoms attached to the carbon atom have different polarizabilities; this then accounts for the dissymmetry of polarizability in the molecule. Such optical activity is said to arise from atomic asymmetry. In example (b) we have three identical (carbon) atoms attached to the asymmetric carbon atom, and therefore there is no atomic asymmetry. The molecule

s optically active, however, because of the conformational asymmetry.
Theoretical studies have shown that optical rotation originates from the
active bonds and not from the asymmetric centers as one might expect.[3]

There is another type of molecule which does not contain asymmetric carbon atoms but nevertheless exhibits optical activity because
it lacks an improper axis.   Examples of these molecules are:

Hexahelicene

Spiro[3.3]heptane

M = Ni^{++}
AA =

OPTICAL ROTATORY DISPERSION

The refractive index of a medium is not a constant but depends on the wavelength.   Thus

$$n = 1 + \frac{k\lambda^2}{\lambda^2 - \lambda_0^2} \tag{13-4}$$

where k and λ_0 are constants for a given medium.   It follows, therefore, that the angle of rotation α shown in Fig. 13-3 will change with the wavelength of the incident polarized light, and we obtain the so-called optical rotatory dispersion (ORD) curves as shown in Fig. 13-4. A represents the plain positive ORD curve, that is, the specific rotation increases with decreasing wavelength, and B represents the plain negative ORD curve. The word *plain* means that there are no maxima or minima in the curve.

CIRCULAR DICHROISM

So far we have discussed only the circular birefringence effect.   In addition to the different speeds of E_L and E_R it is also possible that these two components are absorbed to different extents.   When this happens the magnitude of E_L is no longer equal to that of E_R, and E will no longer oscillate along a single line.   In fact, E will trace out an ellipse as shown in Fig. 13-5.   The medium is said to exhibit *circular dichroism* (CD) and the transmitted light becomes elliptically polarized.

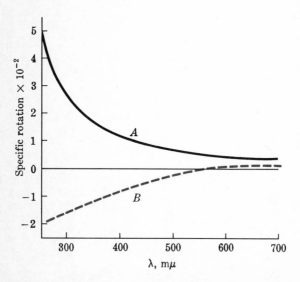

Fig. 13-4 ORD curves. [*By permission from C. Djerassi, "Optical Rotatory Dispersion," McGraw-Hill Book Company, New York, 1960.*]

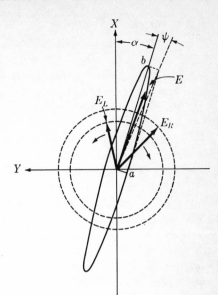

Fig. 13-5 Variation of E for circular dichroism. [*By permission of P. Crabbé, "Optical Rotatory Dispersion and Circular Dichroism in Organic Chemistry," Holden-Day, Inc., Publisher, San Francisco, 1965.*]

In absorption spectroscopy the reduced intensity I is related to the intensity of the incident radiation I_0 by the equation

$$I = I_0 e^{-kl} \qquad\qquad (13\text{-}5)$$

where l is the distance traveled in the medium and k is the absorption coefficient.† The angle of ellipticity Ψ is given by

$$\Psi = \frac{\pi}{\lambda}(k_L - k_R) \qquad\qquad (13\text{-}6)$$

where k_L and k_R are the absorption coefficients for E_L and E_R. Analogous to Eq. (13-2) we write the specific ellipticity $[\Psi]$ as

$$[\Psi] = \frac{\Psi}{lc} \qquad\qquad (13\text{-}7)$$

where l is the path length in centimeters and the molecular ellipticity per unit length $[\Theta]$ is

$$[\Theta] = \frac{[\Psi]M}{100}$$

which can be given in terms of the extinction coefficients ϵ_L and ϵ_R as[4]

$$[\Theta] \simeq 3.3 \times 10^2 (\epsilon_L - \epsilon_R) \qquad\qquad (13\text{-}8)$$

† k is related to the extinction coefficient ϵ by the equation $k = 2.303\epsilon c$ where c is the concentration in moles per liter.

THE COTTON EFFECT

Any medium that exhibits circular birefringence may also exhibit circular dichroism. The combination of these two effects in the region in which the optically active absorption bands are observed gives rise to the phenomenon known as the Cotton effect. Figure 13-6 shows the ORD plots and the corresponding absorption curves for the positive and negative Cotton effects.

Optically active bands are absorption bands of the chromophores which are either intrinsically asymmetric or which become asymmetric as a result of the interaction with the asymmetric environment. As an example of the former we cite the hexahelicene molecule. In this case the *entire* molecule acts as one big chromophore. As an example of the latter we consider the carbonyl group which is symmetric but becomes optically active in an asymmetric environment. Thus the carbonyl group in acetone is optically inactive since it is in a symmetric molecular environment. The same carbonyl group in 3-methylcyclohexanone, however, becomes optically active because there is an asymmetric carbon atom present. We might expect that such induced optical activity would be appreciably smaller than the case in which the whole molecule acts as a chromophore. This is indeed true in general.

Both the absorption and circular dichroism phenomena have their origin in the charge displacements caused by the interaction with the electromagnetic radiation. As a result there will be the induced electric and magnetic dipoles. We define the rotational strength R_k of a chromophore for the k electronic transition as

$$R_k = \rho\mu \cos \theta \tag{13-9}$$

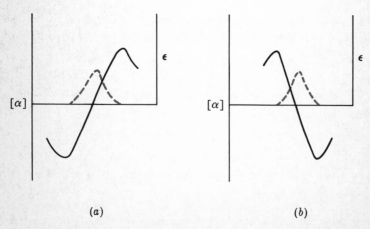

(a)   (b)

Fig. 13-6 ORD plots and the corresponding absorption curves for the positive (a) and negative (b) Cotton effects.

where ρ and μ are the electric and magnetic transition moments and θ is the angle between the directions of these two moments.   For a molecule possessing either a center of inversion or a reflection plane of symmetry, we can have any one of the following three situations:

$\rho \neq 0$ but $\mu = 0$ electric dipole allowed but magnetic dipole forbidden transition

$\mu \neq 0$ but $\rho = 0$ magnetic dipole allowed but electric dipole forbidden transition

$\rho \perp \mu$ $\cos \theta = 0$

In all cases $R_k = 0$ and the molecule is optically inactive.

THE OCTANT RULE

This is a very useful empirical rule in predicting the sign and magnitude of the Cotton effects.[5]   This rule applies only to the substituted cyclohexanones.†   The cyclohexanone molecule is divided into eight octants by three planes A, B, and C (Fig. 13-7).   Plane A passes through carbon atoms 1 and 4; the substituents attached to carbon atom 4 thus lie in this plane.   Plane B encompasses carbon atoms 1, $L2$ and $R2$ where L and R denote left and right from the observer's point of view.   The

† The Cotton effect in this case is associated with the $\pi^* \leftarrow n$ transition of the carbonyl group.

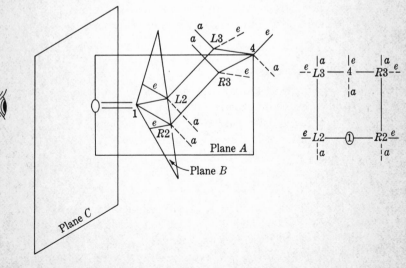

Fig. 13-7 Diagram showing the octant rule. [*By permission of C. Djerassi, "Optical Rotatory Dispersion," McGraw-Hill Book Company, New York, 1960.*]

substituents in the equatorial positions e at $L2$ and $R2$ are practically in this plane. Plane C bisects the carbonyl group and is perpendicular to planes A and B. Thus planes A and B produce four octants and plane C produces four more. The midpoint of the C=O bond is chosen to be the origin of the coordinate system. The octant rule states that substituents in the lower left and far upper right octants make a negative contribution to the Cotton effect; substituents in the far lower right and far upper left make a positive contribution; substituents in any of the three planes make no contribution. Since rarely do substituents bend over the carbonyl group toward the oxygen atom and beyond, the four octants in front of plane C are usually vacant and will not concern us. Thus we shall consider only the four octants defined by planes A and B. As a simple illustration of this rule we consider 3-methylcyclohexanone molecule

If one orients the ketone appropriately for the octant rule to apply, there will be two alternatives. (1) Equatorial representation with the methyl group on the upper left (at L_3) or (2) axial representation with the methyl group on the upper right (at R_3). The fact that a positive Cotton effect is observed confirms that (1) represents the correct conformation. There is no contribution from other groups since carbon atoms 2, 4, and 6 lie in the planes A and B and the contribution of carbon atom 3 is canceled by that of 5.

THE FARADAY AND KERR EFFECTS

These two effects are related to the discussion on optical activity.

When an optically inactive medium is placed in a magnetic field and a beam of plane polarized light is passed through the medium in the same direction of the field, the emerging light will be rotated by an angle ϕ given by

$$\phi = VBl \tag{13-10}$$

where B is the magnetic induction, l is the path length, and V is a constant (for a given medium) called the Verdet constant. This phenomenon is known as the Faraday effect or magneto-optical rotatory dispersion (MORD). In addition to this there is also magnetocircular dichroism (MCD).[6]

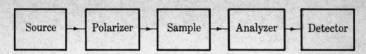

Fig. 13-8 Schematic diagram of a spectropolarimeter.

When an optically isotropic substance is placed in an electric field E, the substance becomes doubly refracting, that is, the refractive index parallel to the direction of the field $n_{\parallel}$ is not equal to that perpendicular to the field $n_{\perp}$. The difference of these two quantities is given by

$$n_{\parallel} - n_{\perp} = KE^2\lambda \qquad (13\text{-}11)$$

This is known as the Kerr electric-optic effect.[7] The quantity K in Eq. (13-11) is the Kerr constant.

13-2 EXPERIMENTAL TECHNIQUES

Figure 13-8 shows the schematic diagram of a spectropolarimeter.† The source and the monochromator are similar to that used in the UV and visible work. The polarizer and analyzer can be Nicol prisms. The detector is a photomultiplier. The intensity of light falling on the detector varies as the analyzer is rotated. The setup for CD measurements is quite similar except that the detecting system is modified to measure the quantity $\epsilon_l - \epsilon_d$.

13-3 APPLICATIONS

Since 1955 the use of ORD in chemistry and biochemistry has steadily increased as a result of the improvements in instrumentation. Its applications can be divided into four areas.

Quantitative analysis Since ORD curves do not exhibit the kind of detailed structure as do the IR or NMR spectra, they are not routinely used in compound identification. However, for a known compound the specific rotation is obviously a good measure of the concentration. For highest sensitivity, $[\alpha]$ (if there is a Cotton effect) is used at either the maxima or minima to compare the unknown with the standard.

Determination of absolute configuration[8] The determination of absolute configuration is of considerable importance in natural product chemistry.

† The term *spectropolarimeter* differs from polarimeter in that in the former case optical rotation can be studied at different wavelengths whereas in the latter case only one wavelength is employed, for example, the sodium D line.

Since the carbonyl group is particularly suited for ORD studies (because of the octant rule) it is fortunate that it is frequently present in many natural product compounds.   Even in its absence it may still be possible to convert other functional groups into the carbonyl group.   For example, cafestol (I), which is found in coffee beans, can be degradated to the corresponding ketone (II) (Fig. 13-9).   A comparison of the ORD curve of

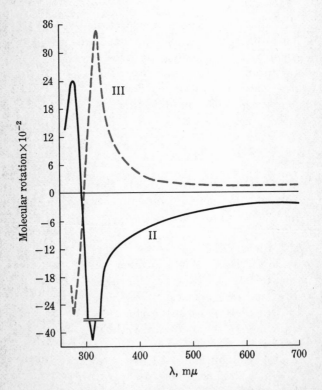

Fig. 13-9 Comparison of the ORD curves for absolute configuration studies. [*By permission of C. Djerassi, "Optical Rotatory Dispersion," McGraw-Hill Book Company, New York, 1960.*]

II with that of 4α-ethylcholestan-3-one (III) shows that these two molecules must have the opposite configuration. Since the absolute configuration of III is known, the configuration of I can therefore be deduced.[9]

The criterion for such studies is simply that there be a compound
with the known absolute configuration for comparison. This compound
must also contain the same chromophore in the same stereochemical and
conformational environment.

Conformation studies   The (+) 3-methylcyclohexanone case discussed in
Sec. 13-1 is a good example of the determination of conformation from

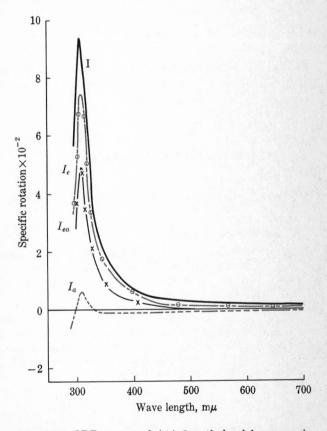

Fig. 13-10 ORD curves of (+) 3-methylcyclohexanone in
methanol (I), methanol-hydrochloric acid (I$_a$), ethanol
(I$_e$), and ethanol-hydrochloric acid (I$_{ea}$). [*From C. Djerassi,
L. A. Mitscher, and B. J. Mitscher, J. Am. Chem. Soc.,* **81:**
*947 (1959). By permission of Dr. Djerassi and the American
Chemical Society.*]

the octant rule. Note that the structure of the compounds must be known in such studies.

Equilibrium studies   When an optically active chromophore takes part in a reaction, the extent of that reaction can sometimes be followed by observing the reduction of the Cotton effect (if there is one). Thus when hydrochloric acid is added to a solution of (+) 3-methylcyclo-hexanone in methanol, the Cotton effect is reduced by 93 percent as a result of the dimethyl ketal formation (Fig. 13-10).

$$\text{>C=O} \rightarrow \text{>C} \begin{matrix} \nearrow OCH_3 \\ \searrow OCH_3 \end{matrix}$$

In ethanol the reduction is only 33 percent, and in isopropyl alcohol there is no reduction.   This affords a convenient procedure for establishing the equilibrium without separation.

SUGGESTIONS FOR FURTHER READING

Introductory

Mislow, K.: "Introduction to Stereochemistry," W. A. Benjamin, Inc., New York, 1965.

Intermediate-advanced

Crabbé, P.: "Optical Rotatory Dispersion and Circular Dichroism in Organic Chemistry," Holden-Day, Inc., Publisher, San Francisco, 1965.
Djerassi, C.: "Optical Rotatory Dispersion," McGraw-Hill Book Company, New York, 1960.
Eliel, E. L.: "Stereochemistry of Carbon Compounds," McGraw-Hill Book Company, New York, 1962.
Mason, S. F.: Optical Rotatory Power, *Quart. Rev.*, **17**: 20 (1963).
Snatzke, G.: Circular Dichroism and Optical Rotatory Dispersion—Principles and Application to the Investigation of the Stereochemistry of Natural Products, *Angew. Chem. Intern. Ed. English*, **7**: 14 (1967).

READING ASSIGNMENTS

Model of Optical Rotation, H. Eyring, *J. Chem. Educ.*, **38**: 601 (1961).
Progress in Our Understanding of Optical Properties of Nucleic Acids, A. M. Lesk, *J. Chem. Educ.*, **46**: 821 (1969).
Circular Dichroism—Theory and Instrumentation, A. Abu-Shumays and J. J. Duffield, *Anal. Chem.*, **38**: 29A (1966).

REFERENCES

1. K. Mislow, "Introduction to Stereochemistry," p. 57. W. A. Benjamin, Inc., New York, 1965.
2. Eliel, p. 401.
3. A. D. Liehr, *J. Phys. Chem.*, **68**: 665 (1964).
4. E. Bunnenberg, C. Djerassi, K. Mislow, and A. Moscowitz, *J. Am. Chem. Soc.*, **84**: 2823 (1962).
5. W. Moffit, A. Moscowitz, R. B. Woodward, W. Klyne, and C. Djerassi, *J. Am. Chem. Soc.*, **83**: 4013 (1961).
6. B. Briat and C. Djerassi, *Science*, **217**: 918 (1968).
7. Ref. 1, p. 190.
8. Eliel, p. 92.
9. C. Djerassi, M. Cais, and L. A. Mitscher, *J. Am. Chem. Soc.*, **81**: 2386 (1959).

14

Masers and Lasers

14-1 GENERAL DISCUSSION

MASERS

Before we discuss the laser we must first mention its predecessor, the maser. Maser is an acronym for *microwave amplification by stimulated emission of radiation*. In Chap. 1 we derived Einstein's relationship between spontaneous and stimulated (or induced) emission [Eq. (1-79)]

$$\frac{A_{nm}}{B_{nm}} = \frac{8\pi h \nu_{nm}^3}{c^3} = \frac{8\pi h}{\lambda_{nm}^3} \tag{14-1}$$

where A_{nm} and B_{nm} are the transition probability per unit time for the spontaneous and stimulated emission between upper state n and lower state m. In the microwave region we have $\lambda \simeq 1$ cm and hence $A_{nm} \ll B_{nm}$. The ratio of the populations in states n and m, N_n/N_m, is given by Boltzmann's expression

$$\frac{N_n}{N_m} = e^{-(E_n - E_m)/kT} \tag{14-2}$$

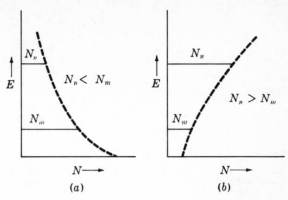

Fig. 14-1 (a) Normal Boltzmann distribution. (b) Population inversion.

From Sec. 1-4 the number of stimulated emissions per second is $N_n B_{nm} \rho(\nu)$ and the number of absorptions per second is $N_m B_{mn} \rho(\nu)$, where $\rho(\nu)$ is the density of the radiation with frequency ν. Since $B_{nm} = B_{mn}$, the number of stimulated upward transitions will always exceed that of downward transitions. However, if by some process we can make $N_n > N_m$, then the number of stimulated emissions will be greater than that of absorption. Such a process is called *population inversion*, which is contrary to the thermal distribution of Boltzmann's law (Fig. 14-1). If such a population-inversion state does exist, then a beam of microwave radiation will *increase* in intensity after interacting with the system.

In 1954 Townes and his coworkers reported the first maser using ammonia.[1] The energy difference between the ground state doublet vibrational level is 0.8 cm^{-1} (see Fig. 7-12). As it turns out, ammonia molecules in these two states have distinctly different electrical properties since they have different electric quadrupole moments. Thus if a beam of ammonia molecules is passed through a strong, inhomogeneous electric field, the two kinds of different (energy) molecules will be deflected differently. Figure 14-2 shows the schematic diagram of the ammonia maser. In this arrangement the upper-state ammonia molecules are focused along the axis of the system, which leads into a microwave cavity whose resonance frequency has been adjusted to 23,870 MHz (0.8 cm^{-1}) while the lower-state molecules escape sideways. If a small amount of microwave power of frequency 23,870 MHz is introduced into the cavity, it will react strongly enough with the molecules in the upper state to make them give up their energy. Thus the output microwave radiation will be greatly enhanced in power.

A number of different types of masers have since been reported.

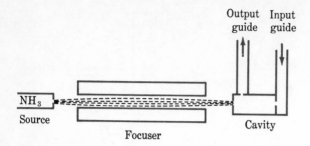

Fig. 14-2 Schematic diagram of the ammonia maser. [*By permission from C. H. Townes and A. L. Schawlow, "Microwave Spectroscopy," McGraw-Hill Book Company, New York, 1955.*]

Although masers have been widely used in satellite communication, radio astronomy, and allied fields, there does not appear to be any direct application to chemistry.

LASERS

Laser is an acronym for *light amplification by stimulated emission of radiation;* it is a maser working at optical frequencies. It was formerly called *optical maser* but is now more commonly referred to as laser.

It is important to distinguish the two terms *chemical laser* and *laser chemistry.* The former is a means of producing lasers by some chemical reactions whereas the latter describes the use of lasers (produced by chemical reactions or physical processes) in chemistry.

A laser consists of three essential components: (1) A system with a set of energy levels of suitable magnitude. (2) A pumping system, that is, a source of energy to excite the system for population inversion. (3) An optical cavity. It is appropriate to divide the methods of producing lasers into the following categories.

PHYSICAL LASERS

Optical pumping method—the ruby laser The first laser produced was that reported by Maiman in 1960.[2] In this arrangement a ruby rod was constructed by doping synthetic sapphire (Al_2O_3) with about 0.05% of Cr_2O_3. Some of the Al^{3+} ions are replaced by Cr^{3+} ions in the crystal lattice. Figure 14-3 shows the energy level diagram for a ruby laser. Transitions from the ground state $^4A_2$ to the excited states $^4F_2$ and $^4F_1$ will occur if the ruby is excited with the appropriate frequencies. The excited states then decay to the ^2E state by radiationless transition. The lifetime of the ^2E state is rather long (about 3 msec at room temperature)

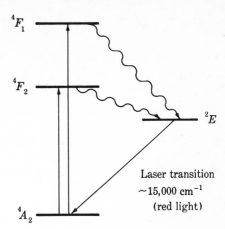

$^4F_1$

$^4F_2$

^2E

Laser transition
$\sim 15{,}000$ cm^{-1}
(red light)

Fig. 14-3 Energy level diagram for a ruby laser.

$^4A_2$

because the $^4A_2 \leftarrow {}^2E$ transition is forbidden by symmetry (spin). Hence the continuous excitation or optical pumping of the system will build up a higher population in the ^2E state compared with the ground state, and there will be a population inversion. Figure 14-4 shows the schematic diagram of a ruby laser. The exciting source is a high-energy xenon lamp which supplies the blue-green excitation for the $^4F_2 \leftarrow {}^4A_2$ and $^4F_1 \leftarrow {}^4A_2$ transitions. The lamp is usually only turned on for a relatively short time—about 1 msec. The emitted radiations ($^4A_2 \leftarrow {}^2E$) along the rod are reflected back through the rod by the two mirrors M_L and M_R (which form the resonant cavity) and thus stimulate more emission from ^2E to $^4A_2$, etc. By making M_R partially transmitting, the laser beam output, which lasts for about 0.5 msec (after optical pumping), can be obtained. The energy of the laser beam ranges from 1 to 100 J at powers reaching 10^5 W cm^{-2}. The power output can be increased further

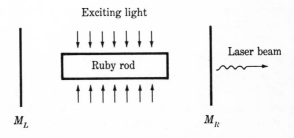

Exciting light

Ruby rod

Laser beam

M_L

M_R

Fig. 14-4 Schematic diagram of the ruby laser.

by delaying the laser action while the system is being optically pumped. This can be achieved by using a fast rotating mirror in place of M_R so that the resonant cavity will be formed only when the two mirrors are parallel to each other. The power of the laser beam output may be optimized by synchronizing the rotation with the flash xenon light output. In this way power as high as 10^9 W cm^{-2} can be obtained.

Collision of the first kind—the argon ion laser   The excited state of an atom or ion X can be more populated compared to the lower states by the inelastic collision between X and an energetic electron. Thus

$$X + e \rightarrow X^* + e$$

In this process some of the kinetic energy of the electron is transferred to X. This method has been applied to produce the neon, argon, krypton, and xenon lasers[3] as well as the argon ion laser.[4]

Collision of the second kind—the helium-neon laser   Consider a two-gas system X and Y in which the metastable state of X lies close to an excited state of Y. Energy transfer from the metastable state of X to the excited state of Y can occur as a result of an inelastic collision. We have

$$X + e \rightarrow X^* + e \qquad \text{collision of the first kind}$$
$$X^* + Y \rightarrow X + Y^* \qquad \text{collision of the second kind}$$

An example of this system is the helium-neon laser in which the helium atom is first excited, by electron impact, to the higher state and then deactivated through collision with the neon atoms. Thus

$$\text{He}(1^1S_0) + e \rightarrow \text{He}(2^1S_0) + e$$
$$\text{He}(1^1S_0) + e \rightarrow \text{He}(2^3S_0) + e$$
$$\text{He}(2^1S_0) + \text{Ne} \rightarrow \text{He} + \text{Ne}^*$$
$$\text{He}(2^3S_0) + \text{Ne} \rightarrow \text{He} + \text{Ne}^*$$

The Ne*'s represent a number of excited levels which lie close to the metastable helium levels. There are a large number of sublevels present, and as a result, over 100 lasers covering the range 0.26 to 133 μm have been observed.

CHEMICAL LASERS[5a,b]

A completely different type of laser, in which the energy released in chemical reactions (other than ionization), is used for pumping. The first chemical laser reported is the gaseous CX_3I system (where X = F or H).[6] When the system is flashed with high-energy radiation the following dissociation takes place:

$$CX_3I \xrightarrow{h\nu} CX_3 + I(5^2P_{\frac{1}{2}})$$

The iodine atom formed is in the excited $5^2P_{\frac{1}{2}}$ state. The population inversion is established between this excited state and the ground state $(5^2P_{\frac{3}{2}})$ with the subsequent laser emission at 13,150 Å.

Another chemical laser was discovered as follows: A mixture of hydrogen and chlorine is flash-photolyzed according to the equations[6]

$$Cl_2 \xrightarrow{h\nu} 2Cl$$
$$Cl + H_2 \rightarrow HCl + H$$
$$H + Cl_2 \rightarrow HCl^* + Cl$$

where HCl* denotes the molecule in its vibrational-rotational excited state. Again population inversion is achieved with the emission of laser light at 37,730 Å.

The chemical lasers can yield very useful information regarding the distribution of energies in a chemical reaction. As more chemical lasers are discovered it seems likely that they will be able to provide detailed information necessary for the understanding of the dynamics of chemical reactions.

PROPERTIES OF LASER LIGHT

Any laser light is characterized by the fact that it is coherent, intense, and monochromatic. The high degree of coherence arises from the fact that the stimulated emission synchronizes the radiation of the individual molecules so that the photon emitted from an excited molecule stimulates another molecule to emit a photon of the same wavelength and exactly in phase with the first photon. The monochromatic nature results from the well-defined energy levels (long lifetimes). For example, the entire output of the ruby laser light is centered at 6943 Å with a width less than 0.1 Å. Although narrow linewidths can also be obtained with monochromators, the intensity of a given laser beam can be about six orders of magnitude greater than a conventional source.

14-2 APPLICATIONS

Laser Raman spectroscopy[7] As we saw in Chap. 8 the principal experimental difficulty in Raman spectroscopy prior to 1964 was the measurement of the extremely weak scattered radiation. Typically, the intensity of Raman scattered lines is about one-hundredth that of Rayleigh lines, the intensity of which is about one-thousandth of the exciting mercury source. However, the use of lasers as the exciting source has greatly enhanced the use of Raman spectroscopy in all branches of chemistry.

Two-photon processes The electronic transitions discussed in Chaps. 9, 10, and 11 occur when the atom or molecule absorbs a photon of the same energy as that between the ground and the excited state. This is the normal one-photon process. However, when a system (atoms or molecules) is irradiated with a high-power laser beam of frequency ν_l, it may undergo a two-photon process, that is, the system goes from the ground to the excited state with the absorption of two photons (Fig. 14-5). The two-photon transitions are of considerable theoretical interest in spectroscopy. The transition dipole moment for the normal one-photon electronic transition is given by $\int\psi_i\mathbf{M}\psi_f\,d\tau$. For all allowed transitions the above integral is nonzero, that is $\psi_i\mathbf{M}\psi_f$ must be an even function. $\mathbf{M}$ depends only on the coordinate and is therefore odd; ψ_i and ψ_f must have different symmetry with respect to each other. For the two-photon process we can imagine the presence of an intermediate state between the initial and final states, that is, transition takes place in two steps. For a two-photon transition to occur the initial and final states must have the *same* symmetry.

It has also been demonstrated that the two-photon processes can lead to chemical reactions which do not ordinarily occur. An example of this is the polymerization of styrene and its derivatives upon irradiation with laser light.[8] The styrene monomer at liquid nitrogen temperature (77°K) is irradiated with a ruby laser beam at 6940 Å. The styrene radicals generated are then polymerized to give polystyrene when the sample is warmed up to room temperature. However, it is known that styrene polymerizes only when irradiated with UV light whose wavelength is about half of that of the laser beam. Thus the most plausible explanation here is that the styrene radicals are generated by a two-photon process.

It seems likely that multiphoton processes will yield useful information regarding the symmetry of the wave functions and also promote chemical reactions not normally attainable by the conventional UV photolysis method.

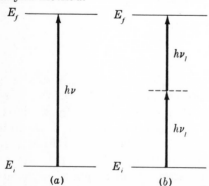

Fig. 14-5 (a) One-photon transition. (b) Two-photon transition.

SUGGESTION FOR FURTHER READING

Intermediate-advanced

Lengyel, O. G. A.: "Lasers," John Wiley & Sons, Inc., New York, 1964.

READING ASSIGNMENTS

The Masers, J. P. Gordon, *Sci. Am.*, **199**(6): 42 (1958).
Optical Masers, A. L. Schawlow, *Sci. Am.*, **204**(6): 52 (1961).
Laser Scattering, B. Chu, *J. Chem. Educ.*, **45**: 224 (1968).
Chemical Lasers, G. C. Pimental, *Sci. Am.*, **214**: 32 (1966).
Laser Chemistry, D. L. Rousseau, *J. Chem. Educ.*, **43**: 566 (1966).
Lasers in Photochemical Kinetics, L. Patterson and G. Porter, *Chem. Britain*, **6**: 246 (1970).

REFERENCES

1. J. P. Gordon, H. J. Zieger, and C. H. Townes, *Phys. Rev.*, **95**: 282 (1955).
2. T. H. Maiman, *Nature*, **187**: 493 (1960).
3. W. B. Bridges, *Appl. Phys. Letters*, **4**: 128 (1964).
4. C. K. N. Patel, W. R. Bennett, Jr., W. L. Faust, and R. A. McFarlane, *Phys. Rev. Letters*, **9**: 506 (1962).
5a. J. V. V. Kasper and G. C. Pimental, *Appl. Phys. Letters*, **5**: 231 (1964).
5b. J. H. Parker and G. C. Pimental, *J. Chem. Phys.*, **43**: 1827 (1965).
6. J. V. V. Kasper and G. C. Pimental, *Phys. Rev. Letters*, **14**: 352 (1965).
7. P. J. Hendra and P. M. Stratton, *Chem. Rev.*, **69**: 325 (1969).
8. Y. Pao and P. M. Rentzepis, *Appl. Phys. Letters*, **12**: 93 (1965).

15

Photoelectron Spectroscopy

15-1 INTRODUCTION

Radiation in the far-UV region ($\lambda < 1700$ Å) can cause electronic excitation, ionization, or even bond rupture in a molecule. However, if we focus our attention only on the ionization process we find that the energy of the ejected electron E is given by the Einstein equation

$$E = h\nu - I \tag{15-1}$$

where $h\nu$ is the energy of the incident radiation and I is the ionization potential, or the binding energy, of the electron. Although Eq. (15-1) has long been employed by the physicists in the study of inner electrons of heavy elements, for example, in β-ray spectroscopy, it has escaped the chemists' attention for some 40 years. Terenin in Russia and Turner in England were the first to realize the applicability of this equation to chemistry.[1]

The basic principle of photoelectron spectroscopy (PES) is quite simple and can be illustrated by the diagram shown in Fig. 15-1. When

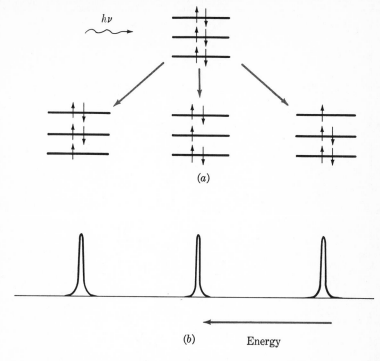

$h\nu$

(a)

(b)   Energy

Fig. 15-1 (a) Photoionization. (b) Energy bands of the ejected electrons.

a photon of energy $h\nu$ is supplied to a molecule, it can remove any one of the valence shell electrons as long as $h\nu$ is greater than the ionization potential. The energy of the ejected electron will of course depend on the particular molecular orbital involved. If the molecular species involved is AB, then this ionization process can be represented as

$$AB \xrightarrow{h\nu} AB^+ + e$$
$$AB \xrightarrow{h\nu} AB^{+*} + e$$

where AB^{+*} denotes the ion in one of its vibrationally excited states (see Sec. 15-3). However, other processes are also possible. Thus when an inner electron is photoejected, an outer electron may fall back to fill the positive "hole," and the energy released can result in the emission of a photon or be dissipated in dissociation. The ionic state initially produced will have a lifetime τ and the uncertainty in energy ΔE ($= \hbar/\tau$), which accounts for the width of the band in the photoelectron spectrum.

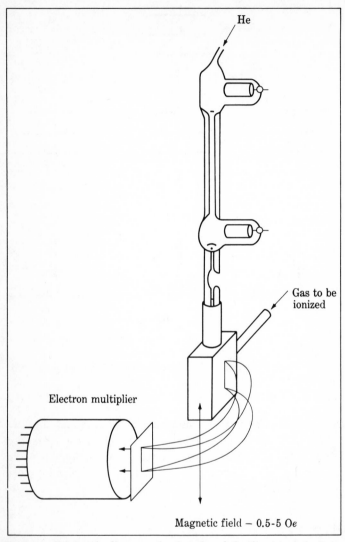

Fig. 15-2 Schematic diagram of a focusing deflection type of photo-electron spectrometer. [*From D. W. Turner, Chem. Britain, **4**: 435 (1968). By permission of Dr. Turner, the Royal Institute of Chemistry, and the Chemical Society.*]

15-2 EXPERIMENTAL TECHNIQUES

In most of the work so far reported a dc discharge tube using helium to provide the monochromatic radiation ($E = h\nu = 21.22$ eV) is employed. Figure 15-2 shows the schematic diagram of a focusing deflection type of photoelectron spectrometer. The compound studied can be in solid, liquid, or vapor form, depending on the pumping techniques.

15-3 THEORY

ATOMIC PHOTOELECTRON SPECTRA

We discuss first the atomic case since it gives the simplest spectra. Figure 15-3 shows the photoelectron spectra of argon, krypton, and xenon excited by the helium resonance radiation. In each case a two-line spectrum with roughly a 2:1 intensity is obtained. This arises as follows. For the argon case, we have

$$\text{Ar} \xrightarrow{h\nu} \text{Ar}^+ + e$$

The term symbol of ground state argon is 3^1S_0; of Ar^+, $3^2P_{\frac{3}{2}}$ and $3^2P_{\frac{1}{2}}$.

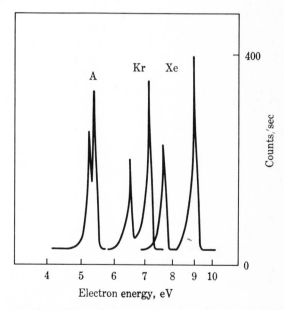

Fig. 15-3 Photoelectron spectra of argon, krypton, and xenon. [*From D. W. Turner and D. P. May, J. Chem. Phys.*, **45:** *471 (1966). By permission of Dr. Turner and the American Institute of Physics.*]

Table 15-1

Rare gas	Ionization potential (eV)		Spectroscopic ionization potential (eV)	
Argon	15 · 79	15 · 93	15 · 755	15 · 933
Krypton	14 · 05	14 · 69	13 · 99	14 · 65
Xenon	12 · 17	13 · 49	12 · 127	13 · 427

SOURCE: M. I. Al-Joboury and D. W. Turner, *J. Chem. Soc.,*
1963: 5141. Reproduced by permission of Dr. Turner and the
Chemical Society.

This accounts for the two-line spectrum and the observed intensity ratio
since the degeneracy of the level is given by $2J + 1$.†

Table 15-1 compares the ionization potentials for the inert gases
obtained by photoelectron spectroscopy with those obtained by the usual
spectroscopic method.

MOLECULAR PHOTOELECTRON SPECTRA

If the photoelectron experiment is carried out at room temperature, we
can safely assume that the molecule is in its ground electronic and vibrational state.   When the molecule is converted into the positive ion it
may end up in an excited vibrational state.   Thus Eq. (15-1) is modified
to give

$$E = h\nu - I_{ele} - E_{vib} - E_{rot} \qquad (15\text{-}2)$$

† The intensity ratio shown in Fig. 15-3, after various corrections, is almost exactly
2:1 in all cases.

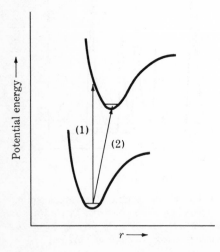

Fig. 15-4 Diagram showing the vertical
(1) and adiabatic (2) transitions.

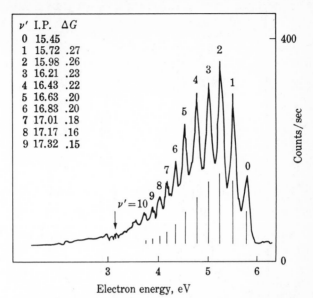

v'	I.P.	ΔG
0	15.45	
1	15.72	.27
2	15.98	.26
3	16.21	.23
4	16.43	.22
5	16.63	.20
6	16.83	.20
7	17.01	.18
8	17.17	.16
9	17.32	.15

Electron energy, eV

Fig. 15-5 Photoelectron spectra of hydrogen. [*From D. W. Turner and D. P. May, J. Chem. Phys.*, **45:** *471 (1966). By permission of Dr. Turner and the American Institute of Physics.*]

where I_{ele} denotes the ionization potential without rotational and vibrational changes. It is appropriate here to distinguish the terms *adiabatic* and *vertical* ionization potential. The former refers to the $v = 0 \rightarrow 0$ transition, which corresponds to the true ionization potential of the molecule, while the latter refers to the most probable transition as predicted by the Franck-Condon principle and therefore usually corresponds to the experimentally measured value. These two transitions are illustrated in Fig. 15-4.

For hydrogen the processes are

1. $H_2 \xrightarrow{h\nu} H_2^+ + e$
2. $H_2 \xrightarrow{h\nu} H_2^{+vib} + e$

where the H_2^+ ion in (1) is in its ground vibrational state and in (2) in the excited vibrational state. Figure 15-5 shows the photoelectron spectrum of the hydrogen molecule excited by the helium resonance radiation. The vertical lines are the calculated Franck-Condon intensities for the transitions

$$H_2^+\ {}^2\Sigma_g^+(v' = 0, 1, 2, \ldots) \leftarrow H_2\ {}^1\Sigma_g^+(v'' = 0)$$

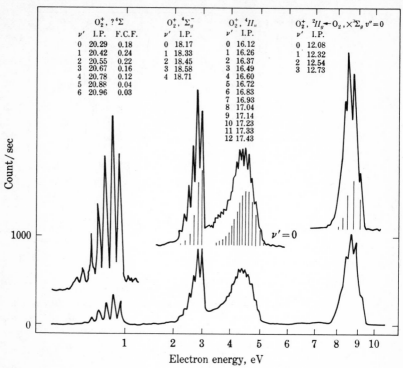

Fig. 15-6 Photoelectron spectrum of oxygen. [*From D. W. Turner and D. P. May, J. Chem. Phys.,* **45**: *471 (1966). By permission of Dr. Turner and the American Institute of Physics.*]

where $^2\Sigma_g{}^+$ and $^1\Sigma_g{}^+$ are the molecular term symbols for $H_2{}^+$ and H_2. Since the lifetime of the ions produced determines the width of the lines (about 0.1 eV in this case), no rotational fine structures can be observed.

The situation for the oxygen molecule is more complex. The electronic configuration of O_2 in its ground state is $KK(\sigma_g 2s)^2(\sigma_u 2s)^2(\sigma_g 2p)^2(\pi_u 2p)^4(\pi_g 2p)^2$ which has the term symbol $^3\Sigma_g{}^-$.[2] The helium resonance

Table 15-2 Molecular orbitals and term symbols of $O_2{}^+$

Molecular orbital from which the electron is removed	Term symbol
$\pi_g 2p$	$^2\Pi_g$
$\pi_u 2p$	$^4\Pi_u,\ ^2\Pi_u$
$\sigma_g 2p$	$^4\Sigma_g,\ ^2\Sigma_g$
$\sigma_u 2s$	$^4\Sigma_u,\ ^2\Sigma_u$

Table 15-3 Vertical π_2 and π_3 ionization potentials of some mono-substituted benzenes

Compound	IP (eV)	$\pi_2 IP - \pi_3 IP$ (eV)
C_6H_6	9·40	
C_6H_5F	9·5, 9·86	∼0·3
C_6H_5Cl	9·31, 9·71	0·40
C_6H_5Br	9·25, 9·78	0·53
C_6H_5I	8·78, 9·75	0·97
$C_6H_5 \cdot CF_3$	9·90	
$C_6H_5 \cdot NO$	9·97	
$C_6H_5 \cdot NO_2$	10·26	
$C_6H_5 \cdot CN$	10·02	∼0·2?
$C_6H_5 \cdot CHO$	9·80	
$C_6H_5 \cdot CH_3$	8·9, 9·13	∼0·2
$C_6H_5 \cdot OH$	8·75, 9·45	0·70
$C_6H_5 \cdot OCH_3$	8·54, 9·37	0·83
$C_6H_5 \cdot OC(CH_3)_3$	8·75, 9·33	0·58
$C_6H_5 \cdot OCF_3$	10·00	
$C_6H_5 \cdot OC_2F_3$	9·97	
$C_6H_5 \cdot NH_2$	8·04, 9·11	1·07
$C_6H_5 \cdot NHCH_3$	7·73, 9·03	1·30
$C_6H_5 \cdot N(CH_3)_2$	7·51, 9·03	1·52
$C_6H_5 \cdot N(C_2H_5)_2$	7·51, 9·11	1·60
$C_6H_5 \cdot N(CF_3)_2$	10·00	

SOURCE: A. D. Baker, D. P. May, and D. W. Turner, *J. Chem. Soc.*, (B) **1968**: 22. Reproduced by permission of Dr. Turner and the Chemical Society.

radiation can remove electrons only from the last four molecular orbitals. Table 15-2 gives the term symbols of O_2^+. Figure 15-6 shows the photo-electron spectrum of oxygen and the various assignments. The vertical lines again represent the calculated Franck-Condon intensities.

In addition to a number of relatively small molecules, photoelectron spectroscopy has also been applied to the aromatic system. For example, there has been considerable theoretical interest in the effect of substituents on the doubly degenerate levels of the benzene molecule.

The two highest bonding molecular orbitals π_2 and π_3 are degenerate. In monosubstituted benzenes this degeneracy is of course removed; the extent of the splitting of π_2 and π_3 for a number of compounds determined by photoelectron spectroscopy is shown in Table 15-3.

15-4 APPLICATIONS

Ionization potentials The PES technique provides perhaps the most direct method of measuring the ionization potentials. It is possible, for example, to obtain the first, second, etc., ionization potentials simultaneously.   The first ionization potential is defined here as the minimum energy required to remove a single electron from the highest filled orbital in a neutral molecule.   The second ionization potential is the energy needed to remove a single electron from the second highest filled orbital in a neutral molecule, etc.   This technique has considerable advantage over the so-called threshold measurements such as the convergence of a Rydberg series, the photoion current method, or the mass spectrometric method,[3] since it is free from the processes of the type

$$AB \xrightarrow{h\nu} A^+ + B^-$$

which make the interpretation of the data difficult.

Nature of the molecular orbital   In a high-resolution spectrum, the vibrational fine structure can often provide useful information regarding the nature of the molecular orbital involved.   As a general rule we note that: (1) The removal of an electron from a nonbonding or very weakly bonding orbital usually results in a single sharp peak without any vibrational fine structure.   The reason is that the equilibrium internuclear distance is not greatly affected by the removal of the weakly bonding electron so that the adiabatic and vertical potentials are practically the same; that is, there are no vibrational changes.   (2) The removal of an electron from a bonding or antibonding orbital usually results in a number of vibrational fine structures.   (3) The removal of an electron from a very strongly bonding or antibonding orbital usually results in a broadband, indicating dissociation or predissociation.   Thus from the appearance of the spectrum and the value of the ionization potentials it is possible to deduce both the nature and the ordering of the molecular orbitals. The limiting factor here lies in the resolution, which is about 0.02 eV at this writing.   The most obvious applications are the checking of theoretical calculations and the assignment of electronic spectra.

Chemical analysis   The PES technique also seems to be well suited to structural and analytical work.   It should be pointed out here that we are not restricted by the helium radiation source.   Indeed, we can equally well employ higher energy sources to study the inner electrons. For example, in the x-ray region the photoelectron spectra show bands of the inner electrons.   Comparison of the spectra of the first-row elements shows that these bands (arising from the K shells) are well separated by 100 eV or so, that is, the bands of Li, Be, B, . . . are far away from each other.   This therefore provides a very convenient means of qualitative analysis and the technique is called electron spectroscopy for chemical analysis (ESCA). A very interesting result of ESCA is that the inner shell electron bands of the *same* element appear at different places if they are bonded to different atoms. Thus the K-shell spectrum of carbon in ethyl trifluoroacetate ($C_2H_5COOCF_3$) consists of four lines, indicating there are four differently bonded carbon atoms present.[4] This is a chemical-shift effect which is analogous to that discussed in Chap. 3. From these chemical shifts and the relative intensities it is possible to make structural assignments in a way analogous to the use of NMR for the same purpose. Similarly, the nitrogen $1s$ spectrum of $Na^+N_3^-$ shows two peaks of relative intensities in a ratio of 2:1, indicating there are two types of nitrogen atoms present. There is little doubt that ESCA will soon be one of the most powerful techniques in chemical analysis.

SUGGESTIONS FOR FURTHER READING

Turner, D. W.: Photoelectron Spectroscopy, *Chem. Britain*, **4:** 435 (1968).
Turner, D. W.: Molecular Photoelectron Spectroscopy, in H. A. O. Hill and P. Day (eds.), "Physical Methods in Advanced Inorganic Chemistry," Interscience Publishers, a division of John Wiley & Sons, Inc., New York, 1968.
Siegbahn, K., et al.: "ESCA Applied to Free Molecules," American Elsevier Publishing Company, New York, 1970.

READING ASSIGNMENTS

Electron Spectroscopy, D. M. Hercules, *Anal. Chem.*, **42:** 20A (1970).
Analytical Potential of Photoelectron Spectroscopy, D. Betteridge and A. D. Baker, *Anal. Chem.*, **42:** 43A (1970).

REFERENCES

1. D. W. Turner and M. I. Al-Joboury, *J. Chem. Phys.*, **37:** 3007 (1962). F. I. Vilesov, B. C. Kurbatov, and A. N. Terenin, *Dokl. Akad. Nauk SSSR*, **138:** ·1329 (1961).
2. G. Herzberg, "Molecular Spectra and Molecular Structure. I. Spectra of Diatomic Molecules," p. 343, D. Van Nostrand Company, Inc., Princeton, N.J., 1950.
3. D. W. Turner, Ionization Potentials, in V. Gold (ed.), "Advances in Physical Organic Chemistry," Academic Press, Inc., New York, 1966.
4. K. Siegbahn et al., *Nova Acta Regiae Soc. Sci. Upsaliensis*, (4) **20** (1967).

16
Summary

We summarize here the conditions for observing spectroscopic resonance for some of the branches discussed in this text in the following chart. Such a chart should serve as a quick reminder; it should also enable one to better appreciate the relationship between different branches, for example, ESR and phosphorescence. Enclosures of similar shape mean that the transition energies are of comparable magnitude. It should be kept in mind that these are only *necessary* conditions.

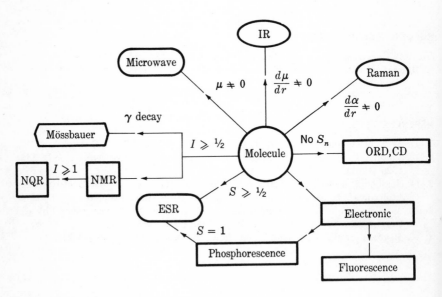

READING ASSIGNMENT

Energy States of Molecules, J. L. Hollenberg, *J. Chem. Educ.*, **47**: 2 (1970).

Matrices

A matrix is a rectangular array of numbers or symbols for numbers. For example

$$\begin{bmatrix} 1 & 6 & 3 \\ 2 & 4 & 5 \\ 7 & 8 & 9 \end{bmatrix}$$

or

$$\begin{bmatrix} a_{11} & a_{12} & a_{13} & a_{14} \\ a_{21} & a_{22} & a_{23} & a_{24} \\ a_{31} & a_{32} & a_{33} & a_{34} \end{bmatrix}$$

In the second case we can represent any element by a_{ij}, where i denotes the columns (the vertical sets) and j denotes the rows (horizontal sets). The dimension of a matrix is defined by the number of columns and rows. Thus a matrix has an $m \times n$ dimension if it has m columns and n rows. A special case is the square matrix in which $m = n$. The trace, or the character (χ), of a square matrix is the sum of the diagonal elements.

For example, for the matrix A

$$\begin{bmatrix} 1 & 7 & 9 \\ 4 & 2 & 5 \\ 8 & 1 & 3 \end{bmatrix}$$

we write $\chi(A) = 1 + 2 + 3 = 6$. A unit matrix is one in which all the diagonal elements are 1 and all the off-diagonal elements are 0.

ADDITION OF MATRICES

Two matrices may be added if they have the same dimensions. Symbolically, this is written as

$$a_{ij} + b_{ij} = c_{ij}$$

As an actual example

$$\begin{bmatrix} 1 & 6 & 3 \\ 2 & 4 & -6 \\ -1 & 4 & -9 \end{bmatrix} + \begin{bmatrix} 2 & 4 & 5 \\ 3 & 6 & 7 \\ 1 & 2 & 8 \end{bmatrix} = \begin{bmatrix} 3 & 10 & 8 \\ 5 & 10 & 1 \\ 0 & 6 & -1 \end{bmatrix}$$

MULTIPLICATION OF MATRICES

Two matrices A and B may be multiplied to give a new matrix C

$$AB = C$$

if A has the same number of columns as B has rows. Thus

$$\begin{bmatrix} a_{11} & a_{12} \\ a_{21} & a_{22} \\ a_{31} & a_{32} \end{bmatrix} \begin{bmatrix} b_{11} & b_{12} \\ b_{21} & b_{22} \end{bmatrix} = \begin{bmatrix} c_{11} & c_{12} \\ c_{21} & c_{22} \\ c_{31} & c_{32} \end{bmatrix}$$

where

$$c_{11} = a_{11}b_{11} + a_{12}b_{21}$$
$$c_{21} = a_{21}b_{11} + a_{22}b_{21}$$
$$c_{31} = a_{31}b_{11} + a_{32}b_{21}$$
$$c_{12} = a_{11}b_{12} + a_{12}b_{22}$$
$$c_{22} = a_{21}b_{12} + a_{22}b_{22}$$
$$c_{32} = a_{31}b_{12} + a_{32}b_{22}$$

THE INVERSE OF A MATRIX

A^{-1} is the inverse of the square matrix A if

$$A^{-1}A = AA^{-1} = E$$

where E is the unit matrix.   For example, if

$$A = \begin{bmatrix} 1 & 2 & 3 \\ 1 & 3 & 3 \\ 1 & 2 & 4 \end{bmatrix} \qquad A^{-1} = \begin{bmatrix} 6 & -2 & -3 \\ -1 & 1 & 0 \\ -1 & 0 & 1 \end{bmatrix}$$

the reader can easily prove that A^{-1} is the inverse of A.

THE TRANSPOSE OF A MATRIX

If we interchange the columns and rows of an $m \times n$ matrix A to obtain
a new matrix A' with dimension $n \times m$, then A' is the transpose of A.
For example

$$\begin{array}{cc} A & A' \\ \begin{bmatrix} 4 & 5 & 1 \\ 2 & 3 & 6 \end{bmatrix} & \begin{bmatrix} 4 & 2 \\ 5 & 3 \\ 1 & 6 \end{bmatrix} \end{array}$$

Character Tables

C_1	E
A	1

C_2	E	C_2		
A	1	1	z, R_z	x^2, y^2, z^2, xy
B	1	-1	x, y, R_x, R_y	yz, xz

C_3	E	C_3	C_3^2		
A	1	1	1	z, R_z	$x^2 + y^2, z^2$
E	1	ϵ	ϵ^*		
	1	ϵ^*	ϵ	$(x,y)(R_x,R_y)$	$(x^2 - y^2, xy)(yz, xz)$

$\epsilon = e^{2\pi i/3}$

C_{2v}	E	C_2	$\sigma_v{}^{xz}$	$\sigma_v{}^{yz}$		
A_1	1	1	1	1	z	x^2, y^2, z^2
A_2	1	1	-1	-1	R_z	xy
B_1	1	-1	1	-1	x, R_y	xz
B_2	1	-1	-1	1	y, R_x	yz

C_{3v}	E	$2C_3$	$3\sigma_v$		
A_1	1	1	1	z	$x^2 + y^2,\ z^2$
A_2	1	1	-1	R_z	
E	2	-1	0	$(x,y)\,(R_x,R_y)$	$(x^2 - y^2,\ xy)\,(xz,yz)$

C_{2h}	E	C_2	i	σ_h		
A_g	1	1	1	1	R_z	$x^2,\ y^2,\ z^2,\ xy$
B_g	1	-1	1	-1	$R_x,\ R_y$	$xz,\ yz$
A_u	1	1	-1	-1	z	
B_u	1	-1	-1	1	$x,\ y$	

C_{3h}	E	C_3	$C_3{}^2$	σ_h	S_3	$S_3{}^5$		
A'	1	1	1	1	1	1	R_z	$x^2 + y^2,\ z^2$
E'	1	ϵ	ϵ^*	1	ϵ	ϵ^*	$x,\ y$	$x^2 - y^2,\ xy$
	1	ϵ^*	ϵ	1	ϵ^*	ϵ	$x,\ y$	$x^2 - y^2,\ xy$
A''	1	1	1	-1	-1	-1	z	
E''	1	ϵ	ϵ^*	-1	$-\epsilon$	$-\epsilon^*$		
	1	ϵ^*	ϵ	$-\epsilon$	$-\epsilon^*$	$-\epsilon$	$R_x,\ R_y$	$xz,\ yz$

$\epsilon = e^{2\pi i/3}$

D_{2h}	E	$C_2{}^z$	$C_2{}^y$	$C_2{}^x$	i	σ^{xy}	σ^{xz}	σ^{yz}		
A_g	1	1	1	1	1	1	1	1		$x^2,\ y^2,\ z^2$
B_{1g}	1	1	-1	-1	1	1	-1	-1	R_z	xy
B_{2g}	1	-1	1	-1	1	-1	1	-1	R_y	xz
B_{3g}	1	-1	-1	1	1	-1	-1	1	R_x	yz
A_u	1	1	1	1	-1	-1	-1	-1		
B_{1u}	1	1	-1	-1	-1	-1	1	1	z	
B_{2u}	1	-1	1	-1	-1	1	-1	1	y	
B_{3u}	1	-1	-1	1	-1	1	1	-1	x	

D_{3h}	E	$2C_3$	$3C_2$	σ_h	$2S_3$	$3\sigma_v$		
A_1	1	1	1	1	1	1		$x^2 + y^2,\ z^2$
A_2	1	1	-1	1	1	-1	R_z	
E	2	-1	0	2	-1	0	$x,\ y$	$x^2 - y^2,\ xy$
A_1	1	1	1	-1	-1	-1		
A_2	1	1	-1	-1	-1	1	z	
E	2	-1	0	-2	1	0	$R_x,\ R_y$	$xz,\ yz$

D_{6h}	E	$2C_6$	$2C_3$	C_2	$3C_2'$	$3C_2''$	i	$2S_3$	$2S_6$	σ_h	$3\sigma_d$	$3\sigma_v$		
A_{1g}	1	1	1	1	1	1	1	1	1	1	1	1		$x^2 + y^2,\, z^2$
A_{2g}	1	1	1	1	-1	-1	1	1	1	1	-1	-1	R_z	
B_{1g}	1	-1	1	-1	1	-1	1	-1	1	-1	1	-1		
B_{2g}	1	-1	1	-1	-1	1	1	-1	1	-1	-1	1		
E_{1g}	2	1	-1	-2	0	0	2	1	-1	-2	0	0	R_x, R_y	$xz,\, yz$
E_{2g}	2	-1	-1	2	0	0	2	-1	-1	2	0	0		$x^2 - y^2,\, xy$
A_{1u}	1	1	1	1	1	1	-1	-1	-1	-1	-1	-1		
A_{2u}	1	1	1	1	-1	-1	-1	-1	-1	-1	1	1	z	
B_{1u}	1	-1	1	-1	1	-1	-1	1	-1	1	-1	1		
B_{2u}	1	-1	1	-1	-1	1	-1	1	-1	1	1	-1		
E_{1u}	2	1	-1	-2	0	0	-2	-1	1	2	0	0	x, y	
E_{2u}	2	-1	-1	2	0	0	-2	1	1	-2	0	0		

$C_{\infty v}$	E	$2C_\infty^\phi$	$\cdots$	$\infty \sigma_v$		
Σ^+	1	1	$\cdots$	1	z	$x^2 + y^2$, z^2
Σ^-	1	1	$\cdots$	-1	R_z	
π	2	$2 \cos \phi$	$\cdots$	0	$(x,y)(R_x, Ry)$	xz, yz
Δ	2	$2 \cos 2\phi$	$\cdots$	0		$x^2 - y^2$, xy
ϕ	2	$2 \cos 3\phi$	$\cdots$	0		

$D_{\infty h}$	E	$2C_\infty^\phi$	$\cdots$	$\infty \sigma_v$	i	$2S_\infty^\phi$	$\cdots$	∞C_2		
Σ_g^+	1	1	$\cdots$	1	1	1	$\cdots$	1		$x^2 + y^2$, z^2
Σ_g^-	1	1	$\cdots$	-1	1	1	$\cdots$	-1	R_z	
π_g	2	$2 \cos \phi$	$\cdots$	0	2	$-2 \cos \phi$	$\cdots$	0	R_x, R_y	xz, yz
Δ_g	2	$2 \cos 2\phi$	$\cdots$	0	2	$2 \cos 2\phi$	$\cdots$	0		$x^2 - y^2$, xy
Σ_u^+	1	1	$\cdots$	1	-1	-1	$\cdots$	-1	z	
Σ_u^-	1	1	$\cdots$	-1	-1	-1	$\cdots$	-1		
π_u	2	$2 \cos \phi$	$\cdots$	0	-2	$2 \cos \phi$	$\cdots$	0	x, y	
Δ_u	2	$2 \cos 2\phi$	$\cdots$	0	-2	$-2 \cos 2\phi$	$\cdots$	0		

Bohr and Nuclear Magnetons

The Bohr and nuclear magneton provide convenient units for the measurement of electron and nuclear magnetic moments. We shall give a simple derivation below. Consider a charge e moving in a circle of radius r. According to classical electricity and magnetism, the magnetic moment μ associated with this motion is

$$\mu = A \times i \qquad\qquad (A3\text{-}1)$$

where A is the area of the circle and i the current. The period of motion T is

$$T = \frac{2\pi r}{v} \qquad\qquad (A3\text{-}2)$$

where v is the velocity of the charged body. The frequency of motion ν is simply

$$\nu = \frac{1}{T} \qquad\qquad (A3\text{-}3)$$

The current i generated is given by

$$i = \frac{e}{Tc} = \frac{e\nu}{c} = \frac{e\nu}{2\pi rc} \tag{A3-4}$$

The magnetic moment then becomes

$$\mu = (\pi r^2)\frac{e\nu}{2\pi rc} = \frac{er\nu}{2c} = \frac{ep}{2mc} \tag{A3-5}$$

where p, the angular momentum, is given by

$$p = mrv \tag{A3-6}$$

Quantum mechanics requires that the angular momentum be quantized. Thus

For electrons: $p = \sqrt{S(S+1)}\,\hbar$ (A3-7)

For nuclei: $p = \sqrt{I(I+1)}\,\hbar$ (A3-8)

We then have

$$\mu_S = -\frac{e\hbar}{2m_ec}\sqrt{S(S+1)} \tag{A3-9}$$

$$\mu_I = \frac{e\hbar}{2M_pc}\sqrt{I(I+1)} \tag{A3-10}$$

The Bohr magneton β is given by

$$\beta = \frac{e\hbar}{2m_ec} \tag{A3-11}$$

and the nuclear magneton β_N

$$\beta_N = \frac{e\hbar}{2M_pc} \tag{A3-12}$$

Equations (A3-9) and (A3-10) are not exactly correct.   Instead, we must write

$$\mu_S = -g\beta\hbar\sqrt{S(S+1)} \tag{A3-13}$$

$$\mu_I = g_N\beta_N\hbar\sqrt{I(I+1)} \tag{A3-14}$$

Where g is the g factor for the electron and g_N is the nuclear g factor. These are *experimentally* determined quantities.

Nuclear magnetic resonance table

Isotope	NMR frequency in MHz for 10-kG field	Natural abundance, %	Relative sensitivity for equal number of nuclei at constant field	Relative sensitivity for equal number of nuclei at constant frequency	Magnetic moment μ in multiples of the nuclear magneton $(eh/4\pi M_p c)$	Spin I in multiples of $h/2\pi$	Electric quadrupole moment Q in multiples of $e \times 10^{-24}\ cm^2$
$\cdot n^1$†	29.165	....	0.322	0.685	-1.9130	$\frac{1}{2}$	
$\cdot H^1$	42.577	99.9844	1.000	1.000	2.79270	$\frac{1}{2}$	
$\cdot H^2$	6.536	1.56×10^{-2}	9.64×10^{-3}	0.409	0.85738	1	2.77×10^{-3}
$\cdot H^3$*	45.414	....	1.21	1.07	2.9788	$\frac{1}{2}$	
$\cdot He^3$	32.434	10^{-5}–10^{-7}	0.443	0.762	-2.1274	$\frac{1}{2}$	
$\cdot Li^6$	6.265	7.43	8.51×10^{-3}	0.392	0.82191	1	4.6×10^{-4}
$\cdot Li^7$	16.547	92.57	0.294	1.94	3.2560	$\frac{3}{2}$	-4.2×10^{-2}
$\cdot Be^9$	5.983	100.	1.39×10^{-2}	0.703	-1.1774	$\frac{3}{2}$	2×10^{-2}
$\cdot B^{10}$	4.575	18.83	1.99×10^{-2}	1.72	1.8006	3	0.111
$\cdot B^{11}$	13.660	81.17	0.165	1.60	2.6880	$\frac{3}{2}$	3.55×10^{-2}
$\cdot C^{13}$	10.705	1.108	1.59×10^{-2}	0.251	0.70216	$\frac{1}{2}$	
$\cdot N^{14}$	3.076	99.635	1.01×10^{-3}	0.193	0.40357	1	2×10^{-2}
$\cdot N^{15}$	4.315	0.365	1.04×10^{-3}	0.101	-0.28304	$\frac{1}{2}$	
$\cdot O^{17}$	5.772	3.7×10^{-2}	2.91×10^{-2}	1.58	-1.8930	$\frac{5}{2}$	-4×10^{-3}
$\cdot F^{19}$	40.055	100.	0.834	0.941	2.6273	$\frac{1}{2}$	
Ne^{21}	....	0.257	....	....	....	$\geq \frac{3}{2}$	
Na^{22}†	4.434	....	1.81×10^{-2}	1.67	1.745	3	
$\cdot Na^{23}$	11.262	100.	9.27×10^{-2}	1.32	2.2161	$\frac{3}{2}$	0.1
$\cdot Mg^{25}$	2.606	10.05	2.68×10^{-2}	0.714	-0.85471	$\frac{5}{2}$	
$\cdot Al^{27}$	11.094	100.	0.207	3.04	3.6385	$\frac{5}{2}$	0.149
$\cdot Si^{29}$	8.460	4.70	7.85×10^{-2}	0.199	-0.55477	$\frac{1}{2}$	
$\cdot P^{31}$	17.235	100.	6.64×10^{-2}	0.405	1.1305	$\frac{1}{2}$	
$\cdot S^{33}$	3.266	0.74	2.26×10^{-3}	0.384	0.64274	$\frac{3}{2}$	-6.4×10^{-2}
S^{35}†	5.08	....	8.50×10^{-3}	0.599	1.00	$\frac{3}{2}$	4.5×10^{-2}
$\cdot Cl^{35}$	4.172	75.4	4.71×10^{-3}	0.490	0.82089	$\frac{3}{2}$	-7.97×10^{-2}
$\cdot Cl^{36}$†	4.893	....	1.21×10^{-2}	0.919	1.2838	2	-1.68×10^{-2}
$\cdot Cl^{37}$	3.472	24.6	2.72×10^{-3}	0.408	0.68329	$\frac{3}{2}$	-6.21×10^{-2}
$\cdot K^{39}$	1.987	93.08	5.08×10^{-4}	0.233	0.39094	$\frac{3}{2}$	
K^{40}†	2.470	1.19×10^{-2}	5.21×10^{-3}	1.55	-1.296	4	
$\cdot K^{41}$	1.092	6.91	8.39×10^{-5}	0.128	0.21453	$\frac{3}{2}$	
$\cdot Ca^{43}$	2.865	0.13	6.39×10^{-2}	1.41	-1.3153	$\frac{7}{2}$	
$\cdot Sc^{45}$	10.343	100.	0.301	5.10	4.7491	$\frac{7}{2}$	
$\cdot Ti^{47}$	2.400	7.75	2.10×10^{-3}	0.659	-0.78712	$\frac{5}{2}$	
$\cdot Ti^{49}$	2.401	5.51	3.76×10^{-3}	1.19	-1.1023	$\frac{7}{2}$	
$\cdot V^{50}$	4.245	0.24	5.53×10^{-2}	5.58	3.3413	6	
$\cdot V^{51}$	11.193	~100.	0.383	5.53	5.1392	$\frac{7}{2}$	0.3
$\cdot Cr^{53}$	2.406	9.54	1.0×10^{-4}	0.29	-0.4735	$\frac{3}{2}$	
$\cdot Mn^{55}$	10.553	100.	0.178	2.89	3.4610	$\frac{5}{2}$	0.5
Fe^{57}	....	2.245	....	....	≤ 0.05		
Co^{57}†	10.0	....	0.274	4.95	4.6	$\frac{7}{2}$	
Co^{58}†	13.3	....	0.25	2.5	3.5	2	

† Indicates radioactive.

Nuclear magnetic resonance table (continued)

Isotope	NMR frequency in MHz for 10-kG field	Natural abundance, %	Relative sensitivity for equal number of nuclei at constant field	at constant frequency	Magnetic moment μ in multiples of the nuclear magneton ($eh/4\pi M_p c$)	Spin I in multiples of $h/2\pi$	Electric quadrupole moment Q in multiples of $e \times 10^{-24}\,cm^2$
·Co⁵⁹	10.103	100.	0.281	4.83	4.6388	7/2	0.5
Co⁶⁰†	4.6	....	5 × 10⁻²	4.3	3.0	5?	
Ni⁶¹	....	1.25	....	....	<0.25		
·Cu⁶³	11.285	69.09	9.38 × 10⁻²	1.33	2.2206	3/2	−0.15
·Cu⁶⁵	12.090	30.91	0.116	1.42	2.3790	3/2	−0.14
·Zn⁶⁷	2.635	4.12	2.86 × 10⁻³	0.730	0.8735	5/2	
·Ga⁶⁹	10.218	60.2	6.93 × 10⁻²	1.201	2.0108	3/2	0.2318
·Ga⁷¹	12.984	39.8	0.142	1.525	2.5549	3/2	0.1461
·Ge⁷³	1.485	7.61	1.40 × 10⁻³	1.15	−0.8768	9/2	−0.2
·As⁷⁵	7.292	100.	2.51 × 10⁻²	0.856	1.4349	3/2	0.3
·Se⁷⁷	8.131	7.50	6.97 × 10⁻³	0.191	0.5333	1/2	
Se⁷⁹†	2.210	....	2.94 × 10⁻³	1.12	−1.015	7/2	0.9
·Br⁷⁹	10.667	50.57	7.86 × 10⁻²	1.26	2.0990	3/2	0.33
·Br⁸¹	11.498	49.43	9.84 × 10⁻²	1.35	2.2626	3/2	0.28
Kr⁸³	1.64	11.55	1.89 × 10⁻³	1.27	−0.968	9/2	0.15
·Rb⁸⁵	4.111	72.8	1.05 × 10⁻²	1.13	1.3483	5/2	0.31
·Rb⁸⁷	13.932	27.2	0.177	1.64	2.7415	3/2	0.15
·Sr⁸⁷	1.845	7.02	2.69 × 10⁻³	1.43	−1.0893	9/2	
·Y⁸⁹	2.086	100.	1.17 × 10⁻⁴	4.90 × 10⁻²	−0.1368	1/2	
Zr⁹¹	4.0	11.23	9.4 × 10⁻³	1.04	−1.3	5/2	
·Nb⁹³	10.407	100.	0.482	8.06	6.1435	9/2	−0.4 ± 0.3
·Mo⁹⁵	2.774	15.78	3.22 × 10⁻³	0.761	−0.9099	5/2	
·Mo⁹⁷	2.833	9.60	3.42 × 10⁻³	0.776	−0.9290	5/2	
·Tc⁹⁹†	9.583	....	0.376	7.43	5.6572	9/2	0.3
Ru⁹⁹	....	12.81	....	....	....	5/2	
Ru¹⁰¹	....	16.98	....	....	....	5/2	
·Rh¹⁰³	1.340	100.	3.12 × 10⁻⁵	3.15 × 10⁻²	−0.0879	1/2	
Pd¹⁰⁵	1.74	22.23	7.79 × 10⁻⁴	0.47	−0.57	5/2	
·Ag¹⁰⁷	1.722	51.35	6.69 × 10⁻⁵	4.03 × 10⁻²	−0.1130	1/2	
·Ag¹⁰⁹	1.981	48.65	1.01 × 10⁻⁴	4.66 × 10⁻²	−0.1299	1/2	
·Cd¹¹¹	9.028	12.86	9.54 × 10⁻³	0.212	−0.5922	1/2	
·Cd¹¹³	9.444	12.34	1.09 × 10⁻²	0.222	−0.6195	1/2	
·In¹¹³	9.310	4.16	0.345	7.22	5.4960	9/2	1.144
·In¹¹⁵†	9.329	95.84	0.348	7.23	5.5072	9/2	1.161
·Sn¹¹⁵	13.22	0.35	3.50 × 10⁻²	0.327	−0.9132	1/2	
·Sn¹¹⁷	15.77	7.67	4.53 × 10⁻²	0.356	−0.9949	1/2	
·Sn¹¹⁹	15.87	8.68	5.18 × 10⁻²	0.373	−1.0409	1/2	
·Sb¹²¹	10.19	57.25	0.160	2.79	3.3417	5/2	−0.8
·Sb¹²³	5.518	42.75	4.57 × 10⁻²	2.72	2.5334	7/2	−1.0
·Te¹²³	11.59	0.89	1.80 × 10⁻²	0.262	−0.7319	1/2	
·Te¹²⁵	13.45	7.03	3.16 × 10⁻²	0.316	−0.8824	1/2	
·I¹²⁷	8.519	100.	9.35 × 10⁻²	2.33	2.7939	5/2	−0.75
·I¹²⁹†	5.669	....	4.96 × 10⁻²	2.80	2.6030	7/2	−0.43
·Xe¹²⁹	11.78	26.24	2.12 × 10⁻²	0.277	−0.7726	1/2	
·Xe¹³¹	3.490	21.24	2.77 × 10⁻³	0.410	0.6868	3/2	−0.12
·Cs¹³³	5.585	100.	4.74 × 10⁻²	2.75	2.5642	7/2	≤0.3
Cs¹³⁴†	5.64	....	6.21 × 10⁻²	3.53	2.96	4	
Cs¹³⁵†	5.94	....	5.70 × 10⁻²	2.94	2.727	7/2	
Cs¹³⁷†	6.19	....	6.44 × 10⁻²	3.05	2.84	7/2	

† Indicates radioactive.

Nuclear magnetic resonance table (continued)

Isotope	NMR frequency in MHz for 10-kG field	Natural abundance, %	Relative sensitivity for equal number of nuclei at constant field	Relative sensitivity for equal number of nuclei at constant frequency	Magnetic moment μ in multiples of the nuclear magneton ($eh/4\pi M_P c$)	Spin I in multiples of $h/2\pi$	Electric quadrupole moment Q in multiples of $e \times 10^{-24}$ cm^2
Ba^{135}	4.25	6.59	4.99×10^{-3}	0.499	0.837	$\frac{3}{2}$	
Ba^{137}	4.76	11.32	6.97×10^{-3}	0.559	0.936	$\frac{3}{2}$	
·La^{138}†	5.617	0.089	9.18×10^{-2}	2.64	3.6844	5	2.7
·La^{139}	6.014	99.911	5.92×10^{-2}	2.97	2.7615	$\frac{7}{2}$	0.9
Ce^{141}†	0.35	....	1.1×10^{-5}	0.17	0.16	$\frac{7}{2}$	
Pr^{141}	11.3	100.	0.234	3.18	3.8	$\frac{5}{2}$	-5.4×10^{-2}
Nd^{143}	2.2	12.20	2.81×10^{-3}	1.07	-1.1	$\frac{7}{2}$	≤ 1.2
Nd^{145}	1.4	8.30	6.70×10^{-4}	0.666	-0.69	$\frac{7}{2}$	≤ 1.2
Sm^{147}	1.47	15.07	8.8×10^{-4}	0.725	-0.68	$\frac{7}{2}$	0.72
Sm^{149}	1.19	13.84	4.7×10^{-4}	0.591	-0.55	$\frac{7}{2}$	0.72
Eu^{151}	10.	47.77	0.168	2.84	3.4	$\frac{5}{2}$	~ 1.2
Eu^{153}	4.6	52.23	1.45×10^{-2}	1.25	1.5	$\frac{5}{2}$	~ 2.5
Gd^{155}	....	14.68	....	....	-0.19	$\frac{3}{2}$	
Gd^{157}	....	15.64	....	....	-0.33	$\frac{3}{2}$	
Tb^{159}	....	100.	....	....	....	$\frac{3}{2}$	
Dy^{161}	....	18.73	....	....	....	$\frac{7}{2}$	
Dy^{163}	....	24.97	....	....	....	$\frac{7}{2}$	
Ho^{165}	....	100.	....	....	....	$\frac{7}{2}$	
Er^{167}	....	22.82	....	....	....	.....	~ 10
Tm^{169}	....	100.	....	....	....	$\frac{1}{2}$	
Yb^{171}	6.9	14.27	4.19×10^{-3}	0.161	0.45	$\frac{1}{2}$	
Yb^{173}	1.98	16.08	1.18×10^{-3}	0.543	-0.65	$\frac{5}{2}$	3.9
Lu^{175}	5.7	97.40	4.94×10^{-2}	2.79	2.6	$\frac{7}{2}$	5.9
Lu^{176}†	....	2.60	....	....	4.2	≥ 7	6–8
Hf^{177}	....	18.39	....	....	....	$\frac{1}{2}$ or $\frac{3}{2}$	
Hf^{179}	....	13.78	....	....	....	$\frac{1}{2}$ or $\frac{3}{2}$	
Ta^{181}	4.6	100.	2.60×10^{-2}	2.26	2.1	$\frac{7}{2}$	6.5
·W^{183}	1.75	14.28	6.98×10^{-5}	4.12	0.115	$\frac{1}{2}$	
·Re^{185}	9.586	37.07	0.133	2.63	3.1437	$\frac{5}{2}$	2.8
·Re^{187}	9.684	62.93	0.137	2.65	3.1760	$\frac{5}{2}$	2.6
·Os^{189}	3.307	16.1	2.24×10^{-3}	0.385	0.6507	$\frac{3}{2}$	2.0
Ir^{191}	0.81	38.5	3.5×10^{-5}	9.5×10^{-2}	0.16	$\frac{3}{2}$	~ 1.2
Ir^{193}	0.86	61.5	4.2×10^{-5}	0.104	0.17	$\frac{3}{2}$	~ 1.0
·Pt^{195}	9.153	33.7	9.94×10^{-3}	0.215	0.6004	$\frac{1}{2}$	
Au^{197}	0.691	100.	2.14×10^{-5}	8.1×10^{-2}	0.136	$\frac{3}{2}$	0.56
·Hg^{199}	7.612	16.86	5.72×10^{-3}	0.179	0.4993	$\frac{1}{2}$	
Hg^{201}	3.08	13.24	1.90×10^{-3}	0.362	-0.607	$\frac{3}{2}$	0.5
·Tl^{203}	24.33	29.52	0.187	0.571	1.5960	$\frac{1}{2}$	
·Tl^{205}	24.57	70.48	0.192	0.577	1.6114	$\frac{1}{2}$	
·Pb^{207}	8.899	21.11	9.13×10^{-3}	0.209	0.5837	$\frac{1}{2}$	
·Bi^{209}	6.842	100.	0.137	5.30	4.0389	$\frac{9}{2}$	-0.4
U^{235}†	....	0.71	....	....	....		
Np^{237}†	~ 20	....	1.0	5.0	6 ± 2.5	$\frac{5}{2}$	
Pu^{239}†	6.1	....	2.9×10^{-3}	0.14	0.4	$\frac{1}{2}$	
Pu^{241}†	4.3	....	1.2×10^{-2}	1.2	1.4	$\frac{5}{2}$	
Free Electron	27,994	....	2.85×10^{8}	658	-1836	$\frac{1}{2}$	

† Indicates radioactive.

SOURCE: Reproduced by permission of Varian Associates.

appendix 5

Relaxation Times

For simplicity let us consider a system containing electrons (or nuclei) which possess a spin of $\frac{1}{2}$. In the presence of a magnetic field of strength H the two levels corresponding to $M_S = \pm\frac{1}{2}$ ($M_I = \pm\frac{1}{2}$) are separated by ΔE and the ratio of the population in the upper state n_u to that in the lower state n_l is given by the Boltzmann expression

$$\frac{n_u}{n_l} = e^{-\Delta E/kT} \tag{A5-1}$$

The total number of spins N is given by

$$N = n_u + n_l \tag{A5-2}$$

and the difference n is

$$n = n_l - n_u \tag{A5-3}$$

Therefore we have

$$n_l = \frac{N + n}{2} \tag{A5-4}$$

$$n_u = \frac{N - n}{2} \tag{A5-5}$$

According to Eq. (A5-1) n_l is always greater than n_u when T is finite. First, let us consider the situation in which the spins do not interact with their surroundings, that is, isolated spins, and radiation with the appropriate frequency is applied to induce transition between the two levels. If $P_{l \to u}$ and $P_{u \to l}$ are the transition probabilities for a spin to make upward and downward transitions, then according to quantum mechanics

$$P_{l \to u} = P_{u \to l} \tag{A5-6}$$

The rate of change of n_l is given by the differential equation

$$\frac{dn_l}{dt} = \frac{1}{2}\frac{dn}{dt} = n_u P_{u \to l} - n_l P_{l \to u} = -Pn \tag{A5-7}$$

where $P = P_{u \to l} = P_{l \to u}$. Equation (A5-7) can be written as

$$\frac{dn}{dt} = -2Pn \tag{A5-8}$$

Solving Eq. (A5-8) we obtain

$$n = n(0)e^{-2Pt} \tag{A5-9}$$

where $n(0)$ is the value of n at $t = 0$. The rate of absorption of radiation energy is

$$\frac{dE}{dt} = P(n_l - n_u)\,\Delta E = Pn\,\Delta E = n(0)P\,\Delta E\,e^{-2Pt} \tag{A5-10}$$

According to Eq. (A5-10) the rate of absorption decreases as t increases, and finally at very large values of t no resonance signals will be observed. Thus

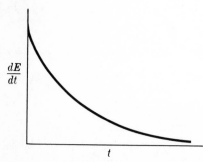

Let us now consider the opposite situation in which there is no radiation but the spins interact with their surroundings. Assume that the interaction gives rise to the upward and downward transitions the probabilities of which are denoted by $P_\uparrow$ and $P_\downarrow$. At thermal equilibrium we write

$$n_l P_\uparrow = n_u P_\downarrow \tag{A5-11}$$

It is obvious here that $P_\uparrow$ and $P_\downarrow$ are not equal since $n_l > n_u$. Again, the rate of change of n_l is given by

$$\frac{dn_l}{dt} = \frac{1}{2}\frac{dn}{dt} = n_u P_\downarrow - n_l P_\uparrow = \frac{N-n}{2} P_\downarrow - \frac{N+n}{2} P_\uparrow$$

$$\frac{dn}{dt} = N(P_\downarrow - P_\uparrow) - n(P_\downarrow + P_\uparrow) \tag{A5-12}$$

We therefore have

$$\frac{dn}{dt} = (P_\downarrow + P_\uparrow)\left[\frac{N(P_\downarrow - P_\uparrow)}{P_\downarrow + P_\uparrow} - n\right] \tag{A5-13}$$

This may be rewritten as

$$\frac{dn}{dt} = -\frac{n - n_0}{T_1} \tag{A5-14}$$

where n_0, the population difference at thermal equilibrium, is given by

$$n_0 = N\frac{P_\downarrow - P_\uparrow}{P_\downarrow + P_\uparrow} \tag{A5-15}$$

and $1/T_1$ is given by

$$\frac{1}{T_1} = P_\downarrow + P_\uparrow \tag{A5-16}$$

T_1 is called the spin-lattice relaxation time.

For the real situation in which there are both radiation induced transitions and interaction between the spins and their environment, we simply combine Eqs. (A5-8) and (A5-14)

$$\frac{dn}{dt} = 2Pn - \frac{n - n_0}{T_1} \tag{A5-17}$$

At equilibrium $dn/dt = 0$ and Eq. (A5-17) gives

$$n = \frac{n_0}{1 + 2PT_1} \tag{A5-18}$$

The rate of absorption of energy at equilibrium is given by

$$\frac{dE}{dt} = nP\,\Delta E = \frac{n_0\,\Delta E P}{1 + 2PT_1} \tag{A5-19}$$

Equation (A5-19) shows that if $2PT_1 \gg 1$ and $2T_1 \gg n_0 \, \Delta E$, no resonance will be observed, and we will then have what is commonly known as the saturation effect. T_1 is a characteristic time for the spin-lattice relaxation mechanism which is the interaction between the spin system with its environment, that is, the lattice.† The spin system is "coupled" to the thermal motion of the lattice and through this coupling magnetic energy can be converted into thermal energy. In this way spins in the higher energy levels can "relax" back to the lower level and the population difference will be maintained. The magnitude of T_1 affects the resonance linewidth as shown by the approximate linewidth parameter equation

$$\frac{1}{T_2'} = \frac{1}{T_1} + \frac{1}{T_2} \tag{A5-20}$$

The interaction among the spins themselves can also lead to line broadenings. This effect, which is called the spin-spin, or transverse, relaxation, arises because each spin experiences a magnetic field due to its neighboring spins, and consequently its energy will vary with time. Therefore, the energy levels corresponding to the upper and lower states are not sharply defined and should be more properly represented as

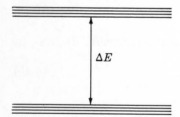

The time required for the energy of the spin to spread over a certain value is given by T_2, the spin-spin, or transverse, relaxation time. The important difference between T_1 and T_2 is that the former involves transitions whereas the latter does not. The values of T_1 and T_2 are usually of the same order of magnitude in solution and the typical values are seconds for the nuclear spin relaxation times and 10^{-5} sec or shorter for the electron spin relaxation times. Finally we note that the linewidth of an ESR or NMR line is predominantly determined by T_1 and T_2; the width due to spontaneous emission is usually negligibly small.

† The word *lattice* usually applies only to solids but in our discussion it includes liquids and gases as well.

The Landé *g* Factor

Consider the following vector model of the individual angular momenta and the total magnetic moment.

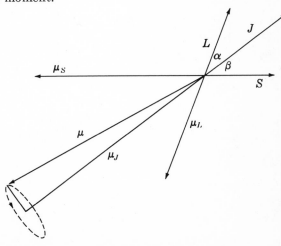

In the Russell-Saunders scheme the resultant magnetic moment μ does not lie along **J** but precesses about **J** as shown.

The magnitudes of **S**, **L**, and **J** are, respectively, $\sqrt{S(S + 1)}\,\hbar$, $\sqrt{L(L + 1)}\,\hbar$, and $\sqrt{J(J + 1)}\,\hbar$. The magnetic moments are given by

$$\mu_S = -g\beta\hbar\,\sqrt{S(S + 1)}$$
$$= -2\beta\hbar\,\sqrt{S(S + 1)} \tag{A6-1}$$
$$\mu_L = -g_L\beta\hbar\,\sqrt{L(L + 1)}$$
$$= -\beta\hbar\,\sqrt{L(L + 1)} \tag{A6-2}$$
$$\mu_J = -g_J\beta\hbar\,\sqrt{J(J + 1)} \tag{A6-3}$$

since

$$\mu_J = \mu_L \cos \alpha + \mu_S \cos \beta \tag{A6-4}$$

we obtain, from Eqs. (A6-1), (A6-2), and (A6-3)

$$g_J = \frac{\sqrt{L(L + 1)}\,\cos \alpha + 2\,\sqrt{S(S + 1)}\,\cos \beta}{\sqrt{J(J + 1)}} \tag{A6-5}$$

From the cosine law,

$$\cos \alpha = \frac{J(J + 1) + L(L + 1) - S(S + 1)}{2\,\sqrt{J(J + 1)}\,\sqrt{L(L + 1)}} \tag{A6-6}$$

$$\cos \beta = \frac{J(J + 1) + S(S + 1) - L(L + 1)}{2\,\sqrt{J(J + 1)}\,\sqrt{S(S + 1)}} \tag{A6-7}$$

Substituting Eqs. (A6-6) and (A6-7) into (A6-5), we obtain

$$g_J = 1 + \frac{J(J + 1) + S(S + 1) - L(L + 1)}{2J(J + 1)} \tag{A6-8}$$

The Born-Oppenheimer Approximation

The mass of the proton ($1.673 \times 10^{-24}\,g$) is about 1840 times that of the electron (0.109×10^{-28} g) so that the electronic motion is much faster than the nuclear motion. If these two motions are assumed to be independent of each other the total wave function ψ of a many-particle (electrons and nuclei) system can be written as a product of the electronic and nuclear wave function

$$\psi = \psi_e \psi_n \qquad (A7\text{-}1)$$

The time-independent Schrödinger equation for the system is given by

$$\left(-\sum_j \frac{\hbar^2}{2M_j} \nabla_j^2 - \sum_i \frac{\hbar^2}{2m_i} \nabla_i^2 + \sum_{j,j'} \frac{Z_j Z_{j'} e^2}{r_{jj'}} - \sum_{i,j} \frac{Z_j e^2}{r_{ij}} \right.$$

$$\left. + \sum_{i,i'} \frac{e^2}{r_{ii'}} \right) \psi = E\psi \qquad (A7\text{-}2)$$

where the first two terms in parentheses denote the nuclear and electronic kinetic energies and the last three terms denote the repulsion between the nuclei, the attraction between the electron and the nucleus, and the repulsion between the electrons, respectively. E is the total energy, and the Z's are the atomic numbers.

In the Born-Oppenheimer approximation we neglect the kinetic energy term of the nuclei when considering the electronic wave equation. We write

$$\left(- \sum_i \frac{\hbar^2}{2m_i} \nabla_i^2 + \sum_{j,j'} \frac{Z_j Z_{j'} e^2}{r_{jj'}} - \sum_{i,j} \frac{Z_j e^2}{r_{ij}} + \sum_{i,i'} \frac{e^2}{r_{ii'}} \right) \psi_e = E \psi_e \qquad (A7\text{-}3)$$

Thus for a given set of nuclear positions ψ_e depends only on the coordinates of the electrons. Of course the positions of the nuclei can and do affect the energy and the wave function of the electron, but for a given set of $r_{jj'}$'s the term

$$\sum_{jj'} \frac{Z_j Z_{j'} e^2}{r_{jj'}}$$

becomes a constant so that the electronic energy is given by

$$E_e = E - \sum_{jj'} \frac{Z_j Z_{j'} e^2}{r_{jj'}} \qquad (A7\text{-}4)$$

Thus Eq. (A7-3) becomes

$$\left[\sum_i \left(- \frac{\hbar^2}{2m_i} \nabla_i^2 - \sum_j \frac{Z_j e^2}{r_{ij}} \right) + \sum_{ii'} \frac{e^2}{r_{ii'}} \right] \psi_e = E_e \psi_e \qquad (A7\text{-}5)$$

The nuclear wave function is given by

$$\left[- \sum_j \frac{\hbar}{2M_j} \nabla_j^2 + E_e(r_{jj'}) \right] \psi_n = E_n \psi_n \qquad (A7\text{-}6)$$

Equations (A7-5) and (A7-6) are the results of the Born-Oppenheimer approximation. We see that in this approximation the total wave function can be separated into an electronic and a nuclear part. Equation (A7-5) is solved for specific values of $r_{jj'}$ and thus E_e depends on the same values. This is illustrated by the familiar plot of E_e versus $r_{jj'}$ for a diatomic molecule. The value of E_e obtained from Eq. (A7-5) is then

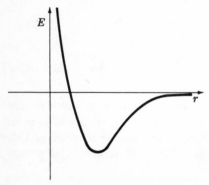

used as the potential energy in Eq. (A7-6).   Note that the solution of Eq. (A7-6) does not depend on the electron coordinates but does depend on the electronic state; hence each electronic state has its own nuclear wave function ψ_n.   The total energy E of the system can be written as a sum of the two separate quantities

$$E = E_e + E_n \tag{A7-7}$$

Index

Index

Absorbance, 24
Absorption photometry, 24
Ammonia, 144, 259
Angstrom unit, 27
Angular wave functions, 11
Anharmonicity, 152
Anomalous Zeeman effect, 196
Argon ion laser, 262
Atomic orbitals, 13
Average value, quantum
    mechanical, 7

Band system, 202
Beer's law, 24
Benzene, 59, 114, 219
Bohr magneton, 106, 284
Boltzmann distribution:
    in electron Zeeman energy
      levels, 107
    in nuclear Zeeman energy
      levels, 55
    in rotational energy levels, 132
    in vibrational energy levels, 151
Born-Oppenheimer approxi-
    mation, 295
Bosons, 45
Butadiene, 45

Center of symmetry, 32
Centrifugal distortion, 134
CH_3Br, 157
C_2H_4, 214
$C_2H_5COOCF_3$, 275
C_2H_5OH, 56, 61, 77
C_6H_6, 59, 114, 219

Character table, 40–42, 280–283
Charge-transfer complex, 88, 228
Chemical shift, 56–60
Chromophores, 225
Circular birefringence, 244
Circular dichroism, 248
CO_2, 156, 176–178
Conjugate variables, 4
Correlation diagrams, 203–205
Cotton effect, 250
Coupling of nuclear spin to
    molecular rotation, 139

de Broglie's equation, 4
Debye-Waller factor, 95
Degrees of freedom, 168
Depolarization ratio, 175
Diatomic molecule, molecular
    treatment of, 202
N,N'-Dimethylacetamide, 76
Dipolar interaction:
    in ESR, 112, 118
    in NMR, 60
Dirac delta function, 112
Dissociation, 207
Doppler effect:
    in linewidth, 25
    in Mössbauer spectroscopy, 93
Double resonance, 74

Einstein's coefficients, 23
Electric field gradient, 81
Electric quadrupole moment, 80
Electromagnetic radiation, 1
Electron spin resonance (ESR)
    spectroscopy, 104–129